LIFE AND DEATH DESIGN
WHAT LIFE-SAVING TECHNOLOGY [...]
EVERYDAY UX DESIGNERS

Katie Swindler

T0248763

NEW YORK   2021

"Katie Swindler carefully blends compelling theory with practical application. Her book is a great deep-dive for anyone interested in understanding how the products and services they build impact behavior, and how to actively design for that responsibility."

—Lauren Liss
Assistant Professor, Columbia College Chicago

"This book provides practical and important advice for designers of all kinds looking to trek beyond the happy path."

—Carolyn Chandler
Coauthor, *A Project Guide to UX Design* and
Adventures in Experience Design

"Swindler's memorable insights are effective for practitioners in all fields; you'll be returning to this one again and again."

—Melissa Smith, PhD
UX Researcher, YouTube

"To gain amazing insight from Swindler's excellent book, you don't need to be someone who designs for people in extremis. You just need to be someone who designs for people under stress, which is, of course, everybody."

—Jon Bloom, PhD
Staff Conversation Designer, Google

Life and Death Design

*What Life-Saving Technology Can Teach
Everyday UX Designers*

By Katie Swindler

Rosenfeld Media, LLC

125 Maiden Lane

New York, New York 10038

USA

On the Web: www.rosenfeldmedia.com

Please send errata to: errata@rosenfeldmedia.com

Publisher: Louis Rosenfeld

Managing Editor: Marta Justak

Interior Layout: Danielle Foster

Cover Design: Heads of State

Illustrators: Shawn Morningstar (line art), Caitlin Everett (image of the brain), Brian Crowley and Wesley Wong (comic book illustrations)

Indexer: Marilyn Augst

Proofreader: Charles Hutchinson

ISBN: 1-933820-84-5

ISBN 13: 978-1-933820-84-2

LCCN: 2021944388

Printed and bound in the United States of America

This book is dedicated to the memory of Pradeep Nayar.
From wizard hats to the Wildbunch,
thank you for showing me many ways to bring joy to work.
You are dearly missed, my friend.

Who Should Read This Book?

This book is for anyone who designs things that are used by people under stress, including:

- Digital and physical products used in fields with life-and-death stakes, like health care or aviation
- Products used in other notoriously high-stress industries, such as day traders or customer service representatives
- Products intended to be used in a moment of high stress, like after a car accident
- Products that shouldn't be stressful to use, but are nevertheless being used by a user in a stressed-out state of mind, such as a mother trying to place an order online while holding a crying baby

Today's user experience professionals are being asked to solve ever-more critical issues across a growing number of sectors. This book gives designers in all fields critical knowledge about human behavior under stress and explores human-centered approaches for designing high-stakes, high-stress experiences.

What's in This Book?

This book contains an in-depth look at everything that designers need to know about the human stress response. It looks at techniques that designers can leverage to harness the benefits of the stress response, promote rational thinking in stressed users, and help users perform to the best of their abilities in critical moments. It also describes how to suppress the negative aspects of stress, avoid panic, and calm users after a crisis.

Chapter 1, "A Designer's Guide to the Human Stress Response," gives an overview of the five steps of the stress response (startle reflex; intuitive assessment; fight, flight, or freeze; reasoned reaction; and recovery), along with a plain-language explanation of the underlying neuroscience. Chapters 2–6 look in detail at the design considerations for each of the five steps. Chapter 7, "Alarms and

Alerts," explores another critical aspect of designing for high-stakes, high-stress experiences: alarms and alerts. And finally, in **Chapter 8**, "Hero by Design," you'll learn about the best ways to inspire users to step up in a moment of crisis and support them as they attempt to intervene and avert disaster.

What Comes with This Book?

This book's companion website (**ⓡrosenfeldmedia.com/books /life-and-death-design/**) contains a blog and additional content. The book's diagrams and other illustrations are available under a Creative Commons license (when possible) for you to download and include in your own presentations. You can find these on Flickr at **www.flickr.com/photos/rosenfeldmedia/sets/**.

What do you mean by "life and death design"? Is this book about death?

Not really. This is a book about high-stakes designs with life-and-death consequences and all the ways your designs can help people in moments of extreme stress or crisis. It's a book about designs that save lives. In particular, **Chapter 8**, "Hero by Design," explores ways to help bring out the best in your users, helping them step up and save the day.

Will I be able to use what's in this book if I design "boring" stuff?

As long as you design something meant to solve a problem for a user, the information in this book will apply to your work. Problems cause stress. Whether someone is under a small or a large amount of stress, the same neurochemicals are released and the same fight-flight-or-freeze instincts drive behaviors. Because of this, lessons learned by designers creating products for extreme environments, like emergency rooms or war zones, can be applied to all sorts of products that help stressed-out users across just about any industry imaginable. In **Chapter 1**, "A Designer's Guide to the Human Stress Response," you'll get an overview of the five phases of the stress response and the design considerations unique to each phase. You'll also get a plain-language overview of the neuroscience that drives these phases.

Does this book cover techniques to address harmful biases?

Absolutely. When humans are stressed, they instinctively fall back on intuition-based decision-making, which has a lot of benefits, but can also open the door for harmful biases to creep in. So **Chapter 3**, "Intuitive Assessment," explores the science of intuition in detail, including how it's formed, when it is most beneficial (and when it's most harmful), and how it can be harnessed through good design. It also looks closely at the role that bias plays in intuitive

decision-making. Then **Chapter 5**, "Reasoned Reaction," explores specific design techniques to help users control bias, even under extremely stressful situations when their instincts might otherwise lead them astray.

I design for a population with people who are chronically stressed. Does this book help me address the unique needs of my users?

The techniques for creating calming designs covered in **Chapter 6**, "Recovery," will be of particular interest for those people designing for users who are chronically stressed. If you are designing for populations with high levels of PTSD, you may also want to review **Chapter 2**, "The Startle Reflex," to learn ways to avoid triggering a startle reflex, as people with PTSD tend to have a particular sensitivity to startling stimuli. **Chapter 7**, which covers "Alarms and Alerts," might also be useful to those designers creating products used in high-stress environments because the chapter discusses techniques to communicate important information appropriately to your users without overwhelming or further stressing them.

CONTENTS

Design is a matter of life and death. It's far too easy for those practicing design to forget the truth of that statement while they are engaged in the practice of it. Sure, we talk at length about making certain that our work is "human-centered," and we throw lots of references to empathy in the mix for good measure, but how far do we really go into the human aspect of human-centricity? Typically, not that far.

I didn't realize to what extent this was true until a few years ago when I took on a design leadership role for a massive, multiyear project to design the upgraded system that child welfare workers used to ensure the delivery of critical services. In the most extreme cases, getting children out of their current situations is a matter of life and death. The workers using this system frequently found themselves under conditions of extreme duress where every minute mattered to ensure a child's safety. My design team and the ones before me all claimed to put the "child at the center" of our work, but what about the worker? It is the fast and thorough response of the person interacting with the system that determines if the necessary interventions happen when needed. The context of their work, their state of mind, and how they as humans respond to stress and imminent danger were all critical factors to be considered. Yet, we barely lifted our heads out of the typical "human-centered design" activities to take those factors into account on anything but the most superficial level.

What we needed was a guide. We needed an accessible way to become well-versed in the human stress response. We needed to fully understand the reflexes, instincts, and intuitive behaviors that affect the humans who would use what we built. We needed this book.

Every chapter of *Life and Death Design* by Katie Swindler is brimming with easy-to-understand explanations of how design impacts critical operations while thoroughly tying the relevant aspects of human biology and psychology to how they shape human behavior in real-world scenarios. Swindler deftly illustrates these concepts with well-researched historical references and relatable true stories—some you know well, and others you may not have heard before. As if that weren't enough, every chapter is heavily cited with a wealth of academic references for further reading.

Life and Death Design not only makes learning complex and necessary subject matter enjoyable, but what you'll learn will also radically change how you approach designing for humans from this point on. Personally, I will be referring to this work and the resources included here again and again, and you should too.

—Lisa Baskett,
Healthcare Design Strategist

It was never my intention to write a book during a global pandemic; however, by the time I first heard the word *coronavirus*, I had already signed the contract. March 15, 2020, was, according to the paperwork, the day I would officially begin writing my manuscript. Instead, it was the day I learned my daughter's school would be closing and my "sweet home," Chicago, would be going into full shelter-in-place lockdown. We were told it would last two weeks. Knowing I had a year to produce the first draft, I figured there would be plenty of time to write after the quarantine period had lifted.

The original concept for the book had been simple—take the fascinating and plentiful research on human stress and design from life-and-death fields like health care, the military, and avionics and see if there were lessons applicable for designers who create products for users who are under other types of "everyday" stress. Little did I know how stressful every single day would soon become.

When it became clear that life was not going to return to normal in two weeks, or really anytime soon, I briefly considered putting the book on hold or canceling it altogether. After all, I had a full-time job as a user experience strategist at Allstate, and my husband and I were suddenly homeschooling my nine-year-old daughter as well. But each day it became more apparent how valuable and applicable knowledge about the human stress response would be for designers, especially digital designers.

Technology had become the glue holding society together, the thing allowing us to maintain our necessary physical isolation while still connecting us in essential ways. In my house, nearly every interaction was being pushed onto digital devices. We worked on devices, played on devices, and escaped the growing sense of existential dread by binge watching *Tiger King* on devices. Even when I woke up in the middle of the night, heart pounding and stomach in knots, it was my phone I would turn to in order to distract myself from my worries until I could fall asleep again.

But the devices had also become a source of stress themselves: the bickering on social media over politics and mask wearing; terrifying stories of racist acts against Asian communities; a seemingly never-ending stream of reports of police brutality and murder, and a groundswell of pain that demanded justice through the Black Lives Matter movement; the bitter election season for the U.S. president that saw five people dead in an attack on our nation's capital; and thousands of people dying each day from a pandemic spiraling out of control. I had said I wanted to write a book that explored the "intersection between stress, technology, and design." Well, now I was living through a symposium on the topic. I realized I couldn't push off my contract and write the book after the pandemic had passed. The more I dug into the research, the more applications I saw for the products and technology I was relying on every day to survive one of the most stress-filled years on record.

All that being said, this is not a book about the coronavirus pandemic. You'll actually only find a couple of brief mentions of it throughout the chapters. But it is a book that was absolutely shaped by the pandemic. It is a book written during a time when the entire world shared a terrible, stressful experience—when we all gained a window of empathy and understanding for those who had experienced even greater levels of trauma. But that understanding will fade over time. In fact, stress makes it more difficult for the brain to turn short-term memories into long-term memories, so the understanding will likely fade quickly. This book was written in a time when the stress was still fresh, the window of empathy still wide open. It is a snapshot capturing insights and getting them down on paper before they became obscured by time and emotional distance.

I believed this was an important book to write long before I heard the word *coronavirus,* but living through a global pandemic has only crystalized why it's so essential. When you design products that people count on, it's critical that they come through for users in a moment of crisis.

When people get stressed, their brains work differently, enhancing some abilities while degrading others. People become more focused, and they make decisions faster. But they also become less rational, less able to cope with quirky interfaces, more error prone, and more in need of support and human connection. If you design your products assuming that people are only using them while in a calm frame of mind, you may inadvertently create a product that fails to support your users in the moment they need it most. My hope is, through this book, designers of all kinds of products will learn how to better empower, support, and protect users in moments of crisis, whether that crisis is small and fleeting or global and years long.

Bias Acknowledgment

Before we dive into the topic in earnest, I want to speak transparently about the potential biases and missing voices within the scientific research behind this book. Much of what we understand today about the human stress response and how stress affects human performance and behavior is informed by foundational research done 40 or more years ago. This research was published almost exclusively by white men. Of the 52 studies and scientific publications I cite in this book, 73% of the authors are male and 92% are white. Exacerbating this skew in perspective is the fact that a large majority of stress research has historically been, and continues to be, funded by military organizations. This means, especially in the early research, the participants often matched the stereotypical soldier profile of the time—young and male.

Being aware of potential skews in the research, I hope, will enable you to think critically about how best to apply the insights in this book to your own work. It may also help you identify areas where doing your own supplemental research is warranted. It is unclear how, or even if, the overrepresentation of white, male researchers and young male participants in early stress studies continues to skew scientists' understanding of the human stress response today. But there are many examples where similar types of skewed data sets have had life-and-death consequences for those who are too dissimilar from the types of people the data favors. For instance, the National Highway Traffic Safety Administration recently called into question the standard calibration and positioning of air bags when it found airbags have severely injured, and even killed, dozens of shorter drivers, mostly women, who might have otherwise walked away with only minor injuries from an accident.

Broadly speaking, the intersection of race, gender, and stress needs more study. While the scientific community seems to be aligned that the basic biological and behavioral aspects of the human stress response are shared across humanity, regardless of gender or race, the research is still unclear about what differences may exist among some of the finer points of the stress response. There are many studies that imply there are differences between male and female reactions to stress, but these studies are so hotly contested within the scientific community that I opted not to include any of the content in this book. I came across no research that spoke to any racial differences in how the stress response works biologically; however, there is a growing body of research around the different *amounts* and *types* of stress experienced by different racial, ethnic, and cultural groups. Those differences impact health, behavior, and the ability to recover from stress. This line of inquiry, primarily focused currently on improving outcomes in areas such as health, academia, and human resources, has the potential to uncover important implications for human factors and design as the research evolves.

Given the known gaps in the scientific research, and aware of the limitations of my own perspective as a white woman living in the United States, I did my best to explore multiple angles, interview diverse sources, and involve many voices from across industries, educational institutes, and cultural backgrounds for this book. In addition to the 52 scientific studies I mentioned previously, I also included another 61 sources made up of designers, authors, industry experts, eyewitnesses, journalists, and technical readers. Although there is significantly better gender and racial diversity among these sources compared to the scientific studies, it is worth noting that they still skew male (55%) and white (70%).

Studying the human condition is a critical component of human-centered design, and as we deepen our understanding of human needs, abilities, and limitations through research, we must ensure that *all* humans are represented in the available studies. This is why diversity within both researchers and research participants is especially critical. This is true not just for visible types of diversity such as sex, race, and age, but also ability, sexual orientation, neurodiversity, religion, culture, and all the other ways that humans are unique. I hope to see significantly more diverse voices from around the world impact future study in both the sciences and design.

Content Warning

This book contains references to topics and world events some readers may find difficult to read, including war, suicide, gun violence, near-death experiences including the near death of children, illness and medical crises, terrorism, and natural disasters.

I have done my best to cover each of these important topics and the people they involve with the respect they deserve. Every story told in this book is shared with an intent to teach designers critical lessons about supporting users through extremely difficult moments. My hope is that by better understanding crises of the past, designers can better support people in the future.

A Designer's Guide to the Human Stress Response

The human stress response is key to our survival as a species. A human without a stress response would never avoid pain, defend themself from an attack, or invent solutions to overcome problems. Stress is unpleasant, but that unpleasantness has a purpose. It's by design. It's meant to drive humans away from harmful things. Stress is a powerful motivator and a major driver of human behavior, which is why it's critical that designers of all types of products, services, and experiences understand the ins and outs of the human stress response.

Stress takes many forms. There's *eustress*, which is a healthy, beneficial type of stress experienced when engaging in enjoyable, challenging experiences. There's also *hypostress*, caused by boredom, that drives the person to take action to relieve the tedium. But when most people think of stress, they think of *distress*, which is a stress response in reaction to negative or overwhelming things. Under the umbrella of distress are the two most common types of stress that scientists study: *acute stress*, stress in response to a momentary crisis that quickly passes, and *chronic stress*, which lasts weeks, months, or years.

Although the causes, severity, and durations of different types of stress responses may vary, at its core, the basic biology of stress is always the same. The stress response has stood the test of evolution for millions of years with the core processes, hormones, and reactions remaining largely unchanged from the time of our earliest mammalian ancestors. In fact, the core difference between the human stress response and the stress response of animals is not a change to the core functionality at all but rather an additional feature—logic. The development of the *prefrontal cortex*, the part of the brain that handles logic and reason, gave humans a way to interrupt, redirect, and suppress their instinctual stress reactions. This additional layer of control means that humans' reaction to stress is significantly more complex than that of other animals, but when you understand the driving forces at the core of the stress response, patterns of behavior emerge. It is through understanding these patterns and better anticipating users' needs in these moments of crisis that designers can make a positive impact.

This book will focus primarily on the acute stress response, because the effect of stress on human behavior is most pronounced in moments of peak stress. The extreme reactions that humans experience during acute stress make it an ideal teaching tool: it's easier to break down, share relevant examples, and establish applicable techniques for designers. And, once you understand how to create designs that support users through an acute stress response, the lessons can be adapted and applied to any other type of stress.

What Happens in a Moment of Acute Stress?

During an acute stress response, the body diverts energy from some physical and mental functions to supercharge others. These changes favor qualities that are associated with "primal" attributes—strength, speed, and aggression. This response sharpens senses, intensifies focus, and drives fast, intuitive decision-making. But these advantages come at a cost: the loss of fine motor control, a suppression of impulse control, a degradation of rational thought and higher-order thinking like reading and math, and a loss of empathy and creativity. The things that people value most about their humanity are momentarily set aside in a raw bid for survival.

Some industries have invested heavily in research around designing for the human stress response. For example, designers who make airplane dashboards, weapon interfaces, and medical devices understand that their users are likely going to be in intense, high-pressure, life-and-death situations when using their products, and they invest in the research necessary to ensure that their products allow humans to function at peak performance while using them. But flying planes, fighting wars, and performing surgeries aren't the only situations where humans experience stress, far from it.

Consider the visceral reaction someone might have to logging in to their banking app to find their account unexpectedly overdrawn, or receiving the devastating news through a social media app that a friend has died. Or think about how often people lose their temper at chat bots or threaten to throw their laptop across the room in frustration over some technical issue.

Of course, digital interfaces aren't just the cause of acute spikes in stress. They are also the place where people often turn for help during a moment of extreme distress—for example, dialing emergency services after a car crash, checking for updates on an approaching storm, frantically Googling for medical information to determine if they should take their child to the ER, or ordering a taxi to escape a date that's turned threatening.

Because technology has become so integrated into every aspect of people's lives, it is a near certainty that, over the course of your career as a digital designer, you will design products that will be used by someone during or directly following a moment of crisis, even if that is not the product's primary intent. By studying and understanding the human stress response, you can better anticipate people's needs and behaviors in those moments, allowing your product to support them in the moments that really matter.

In this chapter, we're going to start with an overview of the acute stress response, which includes five major stages:

1. Startle reflex
2. Intuitive assessment
3. Fight, flight, or freeze
4. Reasoned reaction
5. Recovery

In order to illustrate all the steps of the response, we'll follow the tale of a woman named Amy (named for the neurological star of the stress response—the amygdala) as she experiences a frightening event. Amy and the following story are fictional, but the science underlying Amy's adventure is as real as it gets.

Amy's Accident

Amy is about to have a very bad morning. She's driving to work—it's a route she's followed a hundred times before. She's driving on the kind of mental autopilot only daily commuters can achieve. Engrossed in her podcast episode, Amy begins to execute a standard maneuver, switching from the center lane to the left lane of a three-lane highway, when an unexpected movement is caught by the very edge of her peripheral vision. This is when the trouble begins.

The unexpected movement triggers the first phase of the acute stress response, the *startle reflex*. Her system is flooded with *adrenaline*, a stress hormone that supercharges the body for survival reactions like fighting or fleeing danger. This unconscious reflex causes her to simultaneously turn her attention toward the threat while moving her body away from it. Her arms jerk the wheel sharply to the right in order to move herself and her car out of the path of the incoming object.

Now that Amy has turned to face the approaching object, she enters the second phase of the response, *intuitive assessment*. She effortlessly and instantly identifies the mystery object as a motorcycle recklessly speeding in the left lane. Intuitively, she assesses the trajectory of the motorcycle compared to the trajectory of her own vehicle and, without the need for any actual math, she correctly calculates that she is no longer in danger of colliding with the driver. However, she is so focused on avoiding the cyclist that she doesn't realize her reflexive motion, fueled by adrenaline, was an overcorrection, sending her car veering into the right lane. BAM!

Amy's front bumper clips the side of a minivan. Her vehicle bounces back, thrown into the center lane. This is the moment when Amy officially enters a full *fight-or-flight response*, the third stage of the acute stress response. This stage is fueled by even more adrenaline along with a healthy dose of *cortisol*, another important stress hormone that increases focus, drives immediate action, and prioritizes fast, intuitive decision-making over logic and reason. She wrestles with the wheel to keep from rebounding into the motorcycle on her left. Supercharged by adrenaline, her foot slams down *hard* on the brake. She hears the squeal of tires from multiple vehicles and sees cars swerving all around as they try to avoid rear-ending her. In her panic, Amy is frozen in her seat. Only her eyes move, darting between her mirrors and windows, as the traffic slows to a crawl around her. Miraculously, no additional crashes occur.

A robotic voice fills the car, "Vehicle crash detected. Connecting to OnStar Emergency." Amy remembers that OnStar crash support service came included with her car purchase and is equal parts embarrassed and relieved when a few seconds later a "real human" comes on the line. "This is Randall with OnStar. Is anyone injured?" This rational question kicks Amy's brain out of the panic mode of fight or flight and into the fourth stage, *reasoned reaction*. She reports that she is uninjured, but she is unsure about the people in the other vehicle. Centered by Randall's clear, step-by-step directions, Amy follows his instructions to pull her badly damaged car to the side of the road and confer with the other driver. Luckily, no one in the van is injured either. Still, Randall offers to send a police officer to the scene to create an accident report for the insurance company and Amy agrees.

By the time Amy parks her car on the shoulder, her heart rate and breathing have almost returned to normal. Amy has now entered the *recovery period*, the fifth and final stage. Since her response to the accident involved very little physical exertion, Amy still has plenty of leftover adrenaline in her system, causing shaking hands and jitteriness. While she waits for the police to arrive, she channels her restless energy into using her phone to do all kinds of things: taking down the other driver's contact information, texting her boss to tell her she'll be late for work, and taking pictures of the damage to her own car and the minivan. She even uses her insurance app to file a claim and order a tow truck for her car, and then orders an Uber for herself to the nearest car rental agency. Although it will be another hour or two before the effects of the adrenaline and cortisol fully wear off, Amy, her car, and all the other people and vehicles involved in the incident will make a full recovery from this stressful morning.

Amy's acute stress response both helped and hurt her during this encounter. First, it helped her reflexively avoid a collision with the motorcycle, but this caused her to overcorrect and hit the minivan. Next, it helped her wrestle the car back under control, though in bringing it to a sudden stop, she nearly caused a pileup. And finally, it took her back to rational thinking in the end, allowing her to execute a flurry of activity in the aftermath of the event.

At each step of the response, her ancient instincts had to work with multiple types of modern interfaces: digital, physical, and voice controlled. Some of those interactions were more successful than others. Different parts of the stress response require different approaches from designs. Understanding those various needs is critical to creating designs that can properly protect and empower a user in a moment of crisis.

Startle Reflex Considerations

When you are designing for a startle response, it's helpful to remember that a *startle* response is a powerful force of nature, allowing humans to respond to danger with lightning-fast reactions. You can attempt to harness it through your designs, but this requires your interface to be in just the right place at just the right time. In Amy's story, her hands were on the steering wheel at the moment she was startled, so she was able to use the wheel to jerk out of the path of the motorcycle. This reaction is exactly why drivers are instructed to keep their hands on the wheel at all times while controlling a vehicle. Physical interfaces like steering wheels, buttons, and knobs tend to be better at capturing these speedy reactions than touchscreens, but there are design lessons that can be borrowed from these analog controls to maximize the responsiveness of all kinds of digital interfaces including touchscreens, which we'll explore more in Chapter 2, "The Startle Reflex."

It's also important for designers to keep in mind that, more often than not, startle reflexes are just a nuisance. Either they are false alarms, or they cause someone to overreact, like Amy careening into the right lane after jerking away from the motorcycle. Designers should always take steps to minimize false startle moves, or, if prevention is impossible, put systems in place to protect users from themselves in these uncontrolled, reactionary moments. We'll dig into specific techniques for accomplishing these goals in Chapter 2.

Intuitive Assessment Considerations

When Amy turned to assess the threat level of the speeding motorcycle, this assessment, like all intuitive knowledge, happened instantaneously in her subconscious through a process of pattern matching. Even though she had never seen that particular vehicle before, she was able to match it to the category of objects she had learned was listed as "motorcycles." Similarly, she was able to predict the motorcycle's trajectory intuitively, based on how she had seen similar vehicles move in the past.

It's important to note that nearly all of the driving maneuvers Amy executed throughout this story were powered by her intuition. She never once stopped to calculate how many degrees to turn her wheel to avoid a collision. All of her interactions came from an intuitive understanding of the car's interface, which she had developed through years of repeated use.

NOTE A DANGEROUS LEARNING CURVE

> It takes time and practice to develop reliable intuition for tasks as complex as driving. This is why the first 18 months of driving are so dangerous for new drivers, with car accidents topping the list of causes of death among American teenagers.

Designing for intuition can be a mixed bag. On the one hand, a truly intuitive interface can make technology feel almost like an extension of the user, allowing the person to focus all of their conscious efforts on the problem they are trying to solve. However, there are times when creating interfaces that rely too heavily on intuitive decision-making, unrestrained by fact checking or logic, can lead users to make hasty decisions that are overly influenced by harmful biases and stereotypes. When you are designing for intuition, it is critical to understand the types of environments where this subconscious ability to match patterns and automate decision-making is helpful, and the kinds of situations where users need to have their bias checked by the systems they use. We'll explore different techniques for maximizing the benefits of intuition and minimizing the drawbacks in Chapter 3, "Intuitive Assessment."

Fight, Flight, or Freeze Considerations

When a fight-or-flight response is triggered, the user's rational mind is no longer in charge. No matter what their original objective was, now survival is the only goal. Often, users forget about technical solutions altogether in this state, falling back on more primal methods of dealing with danger. But, occasionally, users are forced through circumstances to interface with technology while in the grips of panic.

During Amy's accident, her use of technology (AKA the car interface) during her fight-or-flight response was highly inconsistent. On the one hand, the fight-or-flight response enhanced her physical strength in a way that helped her get the car steering back under control after hitting the minivan. But the instinct to slam hard on her brakes, a form of the freeze response, actually increased her risk of a second collision with the cars behind her. This kind of unpredictable performance is very common when someone is panicking.

The best thing that technology can do when someone is in fight-or-flight mode is to protect them from harm and get them back to a rational state of mind as quickly as possible. For someone triggered to flee a situation, always allow them to exit or quit if they wish. Additionally, consider ways to provide clear, unobstructed paths to help, preferably human help. For someone in fight mode, look for ways to deescalate the situation. For those who are frozen in fright, give clear, specific direction to help them snap out of their indecision. (For example, the type of direction the OnStar operator Randall provided for Amy.) All of these techniques and more will be explored in Chapter 4, "Fight, Flight, or Freeze."

Reasoned Reaction Considerations

More often than not, to survive and thrive in the modern world requires more logic than instincts. Well-designed systems and services can help users act rationally in a stressful situation by taking complex, multifaceted procedures and breaking them down into step-by-step processes. As illustrated in Amy's story, companies like OnStar will design talk paths for their operators that allow them to triage the information-gathering process quickly and efficiently in an emergency. The goal is to ask about injuries first, and then address safety concerns like getting out of traffic, before moving on to more mundane issues like accident reports and insurance claims. Having these conversation flows written out and streamlined by the design team ahead of time allows the operator to stay calm and focus on the needs of the person in the crash, while executing each step correctly and in the right order of priority.

Experiences that are well crafted for reasoned response help users focus on the most relevant information for the task at hand and make well-informed choices at every step in the process. Techniques for designs that support rational decision-making under stress are explored thoroughly in Chapter 5, "Reasoned Reaction."

Your senses capture raw data from the outside world just the same way a microphone, video camera, thermometer, accelerometer, or other electronic sensor might. And just like a machine, incoming data from each of your senses are turned into electric signals that travel through the brain for processing. When an acute stress response is triggered by something the senses can see, hear, or feel, that signal takes a very specific path through your brain and body, which is mapped in Figure 1.1.

FIGURE 1.1

A map of the signals related to the acute stress response as they travel through the brain and body.

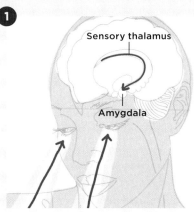

SENSORY THALAMUS: The sensory thalamus is essentially the switchboard for the senses. Normally, it sorts incoming information from the senses and passes it on to the appropriate parts of the brain for decoding. But when it detects something sudden and unexpected, it sends out a special super-fast signal to the amygdala.

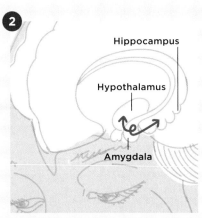

AMYGDALA: The amygdala is the central coordinator within the sympathetic system, the body system that controls the fear response. It's located in the mid-brain sitting just above the spinal cord. When the amygdala receives the emergency signal from the sensory thalamus, it sends two signals, one to the hypothalamus and one to the hippocampus.

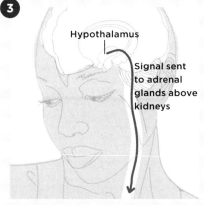

HYPOTHALAMUS: The hypothalamus is a small but mighty region at the base of the brain that controls a number of functions; the most critical to the stress response is the release of the hormone adrenaline that triggers the startle reflex. All of this happens within about 100 milliseconds (a 10th of a second) from when the motion was caught by the peripheral vision.

4

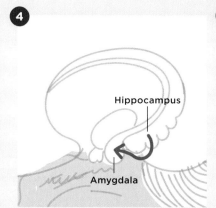

HIPPOCAMPUS: The hippocampus holds experiential memories, for example, memories that are gained through seeing and experiencing things, as opposed to facts or concepts learned in other ways. If the suspected threat matches a memory of something dangerous, then the amygdala will trigger a full fight-or-flight response by activating the HPA axis.

5

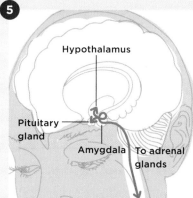

HPA AXIS: HPA stands for hypothalamus, pituitary gland, and adrenal gland, which, when triggered, flood the bloodstream with the stress hormones adrenaline and cortisol. When this happens, energy is redirected from nonessential systems like digestion, reproduction, and immunity to supercharge the circulatory and respiratory systems—the moment when the fight-or-flight reaction truly sets in.

6

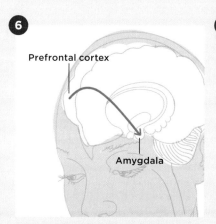

PREFRONTAL CORTEX: Logic and reason come from the part of the brain at the very front of the forehead called the *prefrontal cortex*. Although its response time is a fraction of a second slower than the hippocampus, it has the ability to overrule the more primal survival responses if it thinks the body is taking the wrong actions.

7

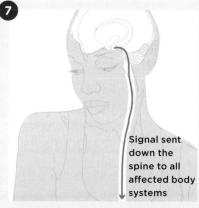

PARASYMPATHETIC SYSTEM: When the crisis is over, the parasympathetic system sends a signal down the spinal cord, telling each body system to return to normal. The circulatory, respiratory, and immune systems bring the heart rate and breathing to normal. The release of acetylcholine restarts any temporarily suppressed processes in the digestive, reproductive, and other nonessential systems.

Recovery Period Considerations

Users often turn to technology for help in the aftermath of a stressful event. In Amy's story, as soon as she resolved the immediate threat, she was on her phone engaged in a flurry of activity.

While the physical aftereffects of adrenaline, like shaking hands, may be of some concern to interface designers, of much more relevance are the lingering mental effects of cortisol during this period. Cortisol drives action. In low-to-moderate levels, it is extremely helpful in terms of focus and motivation. During an acute stress response, however, cortisol floods the brain, intensifying focus to the point of tunnel vision and compelling a person to address the immediate threat to the exclusion of all other goals. High levels of cortisol can cause poor decision-making, reduce creative problem solving, and increase aggression in users. And while the effects of an adrenaline rush rarely last for more than an hour, cortisol can take several hours to be filtered out of the bloodstream.

These negative mental effects have major implications for all kinds of designs used in the aftermath of a stressful event. Luckily, designers can help shorten this recovery period by designing aesthetics that calm the person and lead to empowering actions. In Chapter 6, "Recovery," we'll examine all the techniques that designers can leverage to help users weather the recovery period.

Finding the Peak Stress Points for Your Users

As a designer, being aware of the potential moments of crisis or stress for your users is critical to designing a supportive experience. Here are some questions to ask yourself or discuss with your colleagues to help you find the moments of peak stress within the user experiences you design:

- What are the circumstances where you must deliver bad news to your user? (Example: Service outage or telling a customer a bill is past due.)
- What are the points where the user may discover bad news that your company/digital product is not necessarily aware of yet? (Example: Their account was hacked.)

- What are some upsetting things that might have happened to someone that would cause them to turn to your company/digital product for help? (Example: Customer in car crash and uses insurance app to order a tow truck.)
- If a bully or domestic abuser were trying to use your app/product to harass or harm one of your users, how might they use your digital product to do so? (Example: Abuser cancels power to victim's apartment through a previously shared utility account.)

For additional insights, it can be helpful to map these stressful moments. These stress points can be added to a high-level strategy document like a *customer journey map* that illustrates all of the stages a customer goes through when interacting with your company. Or, if your product is already designed, you can map them directly to your interface designs. Are there certain screens that handle a lot of stressful moments? Or if you are designing voice-based interfaces or diagraming service experiences, are there places within your flow of user interactions that are hot spots for stress? These can be the first places you start to apply the lessons in this book.

It can also be helpful to interview users who have gone through the types of stressful situations you identify and ask them about the type of help they wish they had had in their moment of crisis. Performing a *contextual inquiry*, which means observing someone as they actually experience and deal with the stressful experience, would generate even richer and more accurate insights—however, many crises are rare, dangerous, emotionally traumatic, or unpredictable, making direct observation impossible or inadvisable. Additionally, it's important to be especially mindful of research ethics when interviewing users about past upsetting experiences. Make sure that you are giving them proper compensation, providing a safe environment, and not asking them to relive their trauma.

Critical Information: The Human Stress Response

An acute stress response is triggered when a human is faced with an immediate threat. When designers understand the specific neurological and physiological effects of the stress response, they are better able to create experiences that support users in critical moments. See the phases of the stress response in Figure 1.2.

PHASES OF THE ACUTE STRESS RESPONSE

	STARTLE REFLEX 1–30 seconds	INTUITIVE ASSESSMENT Less than 1 second
DESCRIPTION	When something unexpected and potentially dangerous is detected, a startle reflex is triggered. The body will move away from the threat while the attention moves toward it.	The subconscious mind instantly assesses if the unexpected object is dangerous, and it triggers the fight, flight, or freeze response if the threat is real. If the startle trigger is not dangerous, recovery begins immediately.
LEVEL OF AGITATION		
BRAIN REGIONS ENGAGED	Sensory thalamus Amygdala Hypothalamus	Hippocampus
HORMONES RELEASED	Adrenaline	

FIGURE 1.2

The five stages of the stress response elicit different emotions, actions, and neurological activity.

FIGHT, FLIGHT, OR FREEZE	REASONED REACTION	RECOVERY
Length determined by crisis	1–30 seconds	1–30 seconds
Person will instinctively run away, flee, or, if facing overwhelming odds with no chance of escape, freeze.	Conscious thought is engaged, and logic and reason are used to determine the next steps. Will continue instinctive response if determined to be helpful or override with new actions. Continues until danger passes.	Once danger is determined to have passed, the heart rate and breathing calm, and cortisol and adrenaline levels return to normal over time.

HPA axis Amygdala	Prefrontal cortex	Parasympathetic system Vagus nerve
Adrenaline and cortisol		Acetylcholine

Go to the Source

"Understanding the Stress Response": A plain-language article covering the basics from Harvard Health Publishing, 2011.

The Stress-Proof Brain: Melanie Greenberg has a detailed description of the stress response in the first chapter.

Dr. Dan Siegel's *Hand Model of the Brain*—**YouTube:** A useful visualization technique for the parts of the brain involved in the stress response, 2017, **https://youtu.be/f-m2YcdMdFw**.

Thinking, Fast and Slow: A fascinating book by Daniel Kahneman about the brain and bias.

CHAPTER 2

The Startle Reflex

Eight-year-old Landon Cunningham was sitting in the stands of a Pittsburgh Pirates baseball game on a sunny Florida day, texting a photo to his mom, when Danny Ortiz stepped up to bat. Trying to hit a tricky pitch, Ortiz made a last-second adjustment that caused him to lose hold of his wooden bat. It went flying into the stands straight for little Landon's head. Luckily, Landon's father, Shaun, was there to save the day. Shaun shot his arm out at the last second, protecting Landon's face and taking the brunt of the force on his forearm. *Pittsburgh Tribune-Review* photographer Christopher Horner captured the moment of impact in a brilliant photo shown in Figure 2.1.

FIGURE 2.1
Shaun's "dad reflexes" were captured in a dramatic photo showing the moment of impact.

Because of his fast action, Shaun had a bruise on his arm, but otherwise he and Landon were OK. It certainly could have been worse. A woman hit by a bat at a different stadium a year earlier had suffered a traumatic brain injury.

"He's a hero," little Landon told reporters afterward, and the internet agreed. The Cunninghams became a viral sensation. Shaun's wife,

Ashley Cunningham, gave credit where credit was due, "Thank God he has those reflexes."

A startle reflex is an ingrained part of our survival mechanism. It has evolved in humans because it allows us to dodge out of the way of danger, survive surprise attacks, and protect our loved ones. However, it can also trigger false alarms at inconvenient times, embarrassing us with outsized reactions, and even robbing us of our ability to reason in critical situations. An ancient reflex that evolved to help us survive as hunter-gatherers is not always helpful in dealing with the types of dangers faced by people living in a modern, technology-saturated society.

As a designer, you may be called upon to protect your users from their outdated instincts. Or, if you're especially clever, you may find ways to leverage the power of a startle reflex in your designs. However, to accomplish either of these, you must first understand how the reflex works. In this chapter, we'll explore these key questions:

- What makes something startling and how can designers prevent or control a startle response in their users?
- What are the pros and cons of a startle reflex and how can designers harness the beneficial aspects?
- How can designers protect users from the downsides of a startle reflex?

Dissecting a Jump Scare

People usually hate being startled. It's scary, you lose control, and if the danger isn't real, you look kind of ridiculous. Unless you are designing user interfaces (UIs) for games and entertainment products, you are probably more interested in learning how to avoid startling your users. However, to eliminate startling elements from your design, first you must understand what causes the reflex. "Know thine enemy," as General Sun Tzu recommends.

The film industry coined the term for purposefully triggering a startle reflex; it's called a *jump scare*, and since the 1920s, editors have been perfecting it as an art and a science. Jacob Metiva, an L.A.-based post-production supervisor who has managed the editing process of many films, shared the trade secrets of a well-constructed jump scare with me, and these editing techniques line up perfectly with scientific research on the startle reflex.

There are four factors that psychologists say contribute to creating a startling experience: novelty, intensity, rise time, and priming. While it's impossible to guarantee a startle reflex (startle responses vary widely between individuals), a careful mix of these elements can scare someone quite reliably. Metiva reveals how each of these factors is leveraged by editors and entertainers to illicit jump scares and keep audiences on their toes.

Novelty

In order for sounds or sights to be startling, they must be *novel*, meaning there is a significant change from what came before, usually something unexpected or unpredictable. This change might be the unexpected boom of a car backfiring or the sight of someone standing behind you when you previously thought you were alone in the room.

In the film industry, relying on a *non sequitur*, an action that does not follow logically from the action before it, is considered by many to be the "cheapest" type of jump scare. As an example, Metiva pointed to the 2012 film, *The Woman in Black*, as a particularly egregious offender: "Daniel Radcliffe can't open a box of crackers without some birds flying out."

The birds might be a cheap trick, but it's still an effective way to startle someone, because their subconscious tends to treat anything unexpected as a threat. At the root, this reaction is the same instinctive mistrust and defensiveness that users experience when a design or flow does not meet their expectations. When a site asks for a Social Security number too soon in the registration process, or if the branding on the credit card page doesn't match the rest of the site, the sudden introduction of an unfamiliar element immediately triggers a defensive posture.

It may not be a full popcorn flipping startle response, but it is grounded in the same reflexive negative response to a startlingly unfamiliar experience.

Intensity

If the *intensity* of a trigger is high enough, even if it's expected, it can still provoke a startle response.

Metiva pointed to an interview with Monty Python's Terry Gilliam who shared that he once had a grand plan to pull a brilliant prank on all the TV viewers of their popular *Flying Circus* sketch comedy

In myths and legends, there are myriad examples of mystical portals and dangerous fey creatures that can only be seen when you aren't looking directly at them. Like many folktales, those stories serve as a metaphor for a scientific truth: your central vision is essentially the eyes of the logical, conscious part of your brain, while your peripheral vision allows your base, subconscious survival mechanisms to keep watch over your surroundings.

Although the peripheral vision may lack the clarity and detail of the central vision, it detects movement much more acutely and has a special shortcut in the brain that can send alerts about incoming threats at speeds faster than conscious thought. This shortcut heightens reflexes to deal with threats approaching from the side, but it also means that movement or unexpected figures caught by the peripheral vision are much more likely to trigger a full startle response. Additionally, participants in a lab study were shown to have the ability to process and react to facial expressions seen only in the peripheral vision, especially expressions of fear and anger, evoking responses like increased heart rate, alertness, and vigilance even when the person wasn't consciously aware that they saw anything amiss.

Understanding the strengths and weaknesses of both central vision and peripheral vision is essential for interface designers, especially those designing for AR experiences or multiple displays, such as a car dash. Animations placed at the edge of a user's vision are actually more likely to pull their focus than those placed at the center, and they'll react to it faster. Don't discount the power held at the edges of perception. What is seen there may not be perfectly clear, but that doesn't mean it is any less useful or important for the humans you're designing for.

show. "One thing I wanted to do at one point was that the show would be going on, and we would slowly take the sound down in a sketch. And do it very slowly so people all over England would then get up, turn up the sound a little bit. Then it would go down a little bit more." He mimed people turning the set up a bit more. "And this would go on for about five minutes until [the TV] was at its maximum level, and then we would make the biggest noise we could and blow out every set in the nation." Sadly (or perhaps wisely), the BBC put the kibosh on the devious comedian's mischief, and the plan never came to fruition. It undoubtedly would have been a legendary simultaneous startle of people across Great Britain, a joke completely in keeping with the irreverent Pythonian humor.

Intensely loud auditory triggers are the most reliable way to trigger a startle. A short, sudden sound of 100–125 decibels (dB) is often used as a startle trigger for scientific purposes, but anything over 80dB when played in a quiet room (~40dB) has a pretty high likelihood of making you jump.

Beyond pure volume, a high-contrast visual change or an extremely scary image can amp up the intensity factor. Video games like *Monster Hunter* use oversized monsters to increase the intensity factor and startle players into a full freak-out. Sizing the creatures to completely fill the field of view of the player maximizes the effect, as you can see in Figure 2.2

FIGURE 2.2

Hilarious reaction videos, posted by Twitch streamers, demonstrate how effectively giant "boss" monsters heighten players' reactions.

Rise Time

Rise time, or how suddenly a stimulus reaches intensity, is another factor for startles. A jet engine starting up can become as loud as a pistol shot, but it doesn't startle you because the intensity of the sound increases gradually over time.

Here is the real secret of a cinematic jump scare: what is shown doesn't have to be scary, it just has to be sudden. Metiva pointed to the title cards used to mark the passage of time in Stanley Kubrick's *The Shining* as a great example of this phenomenon (shown in Figure 2.3). "They are so abrupt, you jump out of your chair, but, like, it's just... the date."

FIGURE 2.3
In this scene from *The Shining*, music rises ominously while the camera shows the characters walking happily through the maze. Suddenly, the screen flashes black with just the single word, "Tuesday." It shouldn't be scary, but it absolutely is.

Rise time holds the true key for a designer wanting to eliminate a startle response. The best way to make something less startling is to have the change happen more gradually. Do the opposite of what Kubrick does—instead of sharp, abrupt changes, use fading effects on visuals, audio cues, or animations. This technique will give any interface a more "chill" vibe and completely eliminate the risk of triggering a distracting startle response.

Designers of alarm systems that need to reach loud volumes to project over long distances often use this kind of fade-in effect to reduce startles. A well-designed alarm noise begins with a short lead-in sound that ramps up in both pitch and volume—think of the rising whine of an American police siren, as shown in Figure 2.4. (Find more about alarm design in Chapter 7, "Alarms and Alerts.")

FIGURE 2.4
The diamond-shaped rise and fall of the sound wave shows there aren't any abrupt changes in the sound that would cause a startle reaction.

There is a rule of thumb in design that "contrast equals drama." Human instincts insist that sudden changes equal danger and therefore must be paid attention to. Designers and artists have been hacking this quirk of human attention for centuries. A designer who wants to add a dramatic flair to a party invite might use a bold black-and-white theme. Similarly, a musician trying to heighten emotional impact might incorporate sudden changes in volume throughout the piece. To bring down the drama, designers need only to add midpoints. A black-and-white theme becomes much calmer with the introduction of several gradually darkening shades of gray, and the song becomes more soothing if the volume changes happen gradually across the length of the piece. The effect of rise time is the underlying mechanic behind why these design tricks work.

Priming

In addition to carefully calibrating novelty, intensity, and rise time, Hollywood does one additional thing to increase the effectiveness of a jump scare—ratchet up the tension. Music like the famous *Jaws* movie theme starts by slowing down the action and then steadily winding up the tension bit by bit.

The more stress that someone is under, the more active and intense their startle response will be. This fact is why you will see an exaggerated startle response in people suffering from chronic anxiety disorders, such as post-traumatic stress. According to Dr. Christian Grillon of the National Institute of Mental Health, "This is not because a slamming door reminds them of their trauma, but because they are chronically anxious and the slamming door makes them startle."

The priming effect isn't just triggered by stress from negative experiences or emotions. Those who are experiencing the more positive eustress while focused on a challenging task are also more susceptible to a startle effect. To demonstrate this, Metiva pointed to an internet meme from the early days of the web, now referred to as the *Scary Maze Game* shown in Figure 2.5.

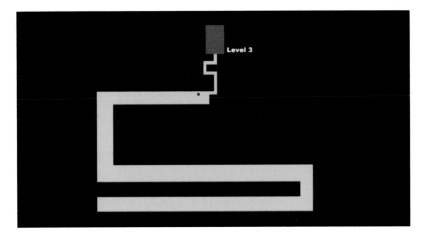

FIGURE 2.5
The third and final level of the *Scary Maze Game* required intense concentration to navigate the tiny path right before the startle was triggered.

A devious developer by the name of Jeremy Winterrowd created one of the earliest and best known of these maze games in 2003. Someone would receive the link to the game, which would lead them through a series of mazes where they had to move their mouse through a progressively narrower path, concentrating more and more intensely in order to finish each level. When the user got to the end point of the third maze, a loud scream played, and the image cut to a terrifying, close-up picture of the woman from *The Exorcist*. If you love watching people wig out, type "Scary Maze Game" into YouTube and enjoy hours of reaction videos of people pranking their friends.

The more intensely your user is focused on a stressor, the more primal their reactions to new challenges will be. Stress intensifies both the positive and negative effects of the startle reflex, as well as other primal responses, like intuition and fight or flight (which we will explore further in the next two chapters). Stress also suppresses a user's logic and reason, making it harder for their rational mind to keep bias and bad habits in check. Often, these high-stress periods are when users most need protection from inevitable human errors, as well as additional paths to recover from a mistake or overreaction. These moments are the place where good design can make the biggest impact.

Harnessing a Startle

Imagine that you're at a track and field competition and a race is about to begin. As the athletes take their marks, pay special attention to the runner in the first lane, who is tense and ready, waiting for the starting pistol to fire. She is about to get an unfair advantage in this contest.

The starting official gets into position, just a few feet away from the runner in the first lane. The official pulls the trigger of the starter pistol, and it erupts with a loud BANG! For the runner, that shot is incredibly loud, more than 100dB. This triggers her startle reflex, bypassing her need to connect the sound of the pistol with its meaning. The startling sound flings her to action 18 milliseconds (ms) faster than the runners farther away who heard the sound at a less startling intensity. Obviously, 18ms is less than 1/50th of a second, but when it comes to a photo finish in an Olympic race, it can be the difference between a silver or gold medal.

This startle-reflex advantage of 18ms on average was discovered during a study done on track and field runners at the University of Alberta in 2005. It's a great example of a design decision conflicting with reflexive human behavior. The race designers intended for a starter pistol to be a neutral signal that could be easily heard by all contestants. But, because they didn't consider the startle reflex, they gave one runner an unfair advantage, undermining the integrity of the contest. Only when this phenomenon was properly understood by the game officials could the design be fixed. Seven years after the study was published, at the 2012 Olympics, the traditional starter pistol was replaced by an electronic beep played simultaneously over multiple speakers, one placed an equal distance behind each runner on the course shown in Figure 2.6.

FIGURE 2.6
A small electronic speaker is built directly into the back of each runner block, such as these blocks from OMEGA, to ensure consistent sound levels for each racer.

Predictable Movements

Biologists, physiologists, neurologists, and psychologists have studied the human startle reflex extensively. The primary benefit is increased response time for defensive movements, and the accompanying adrenaline release superpowers people's muscles for big moves like running away or punching an attacker. Scientists have done studies recording down to the millisecond how fast each major muscle group can respond when triggered. It's easy to chart the path of the signal down the spinal cord as you look at Figure 2.7, which illustrates commonly cited average human response time data published by researcher Michael Davis in 1984.

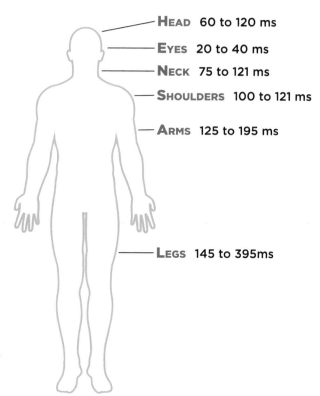

HEAD 60 to 120 ms

EYES 20 to 40 ms

NECK 75 to 121 ms

SHOULDERS 100 to 121 ms

ARMS 125 to 195 ms

LEGS 145 to 395ms

FIGURE 2.7
The average range of human response times of a startle from eyes to legs as published in *The Mammalian Startle Response* by neuroscientist Michael Davis.

During a startle response, instinctive movements are focused on protection first, with threat assessment as a close second. If you're trying to predict the movement of a startled user, it's a safe bet that their body will jerk away from the perceived threat, usually rounding the shoulders and twisting to protect center mass while ducking their head. What's interesting about head movement is, even as they duck their head and move away, more likely than not, their face will instinctually turn *toward* the threat in order to keep it in sight. This instinct happens due to the criticality of threat assessment as part of the reflex. Even if the user is able to suppress their physical startle reaction (which can be done through practice), there is nothing that can prevent their mental attention from shifting to the new threat.

Another predictable reaction for designers to consider is that users will often drop whatever they are carrying during a startle response. Exact arm and hand movements are hard to predict, but the two most likely scenarios are to clench the hands into fists and draw them into the body to protect the head or internal organs, or to spread the fingers and move the arms to brace against an incoming blow. (Flip back to Figure 2.1 and look closely at the body positions of all the different people in the photo to see real examples of these two types of instinctive body positions.) The second reaction, spreading the fingers, is one of the reasons it's necessary to include wrist tethers on VR/AR controllers, because this involuntary reaction can cause them to go flying. If you design handheld devices that are used in immersive games or other environments with a high likelihood of a startle reflex, then you might consider adding tethers to your devices as well.

Gross- vs. Fine-Motor Benefits

The startle reflex evolved to allow people to jump-start their fight-or-flight response. Therefore, it gives a big boost to gross-motor tasks, super powering people's punches and letting them run like the wind. But, in today's modern world, a large portion of the workforce spends the majority of their time each day behind a computer screen. Will the startle reflex help them deal with a crisis in their cubical, tapping into their super speed to meet a deadline?

Unfortunately, no. The startle effect advantage seen when testing the movement of large muscle groups like arms and legs disappears almost completely when looking at the fine motor movement of individual fingers. A 1990 lab study, which involved strapping a lot of fingers and arms to tables, showed that although intentional *arm* movement has a significant improvement in speed when participants are startled, there is no statistical improvement in intentional *finger* movements under the same conditions.

Other studies have shown that fingers become stiff and clumsy under the effects of a startle, reducing the ability to accurately draw even simple lines. This effect means that, as a digital designer, you sadly can't count on the startle reflex to help at a typical keyboard, mouse, or touchscreen.

Most touchscreen designers follow established best practices to ensure that their products are usable by people with a wide range of finger sizes, sometimes referred to as *fat finger design*. For instance, Android's "Material Design Guidelines" specify that any *touch target*, or clickable element, should be 7–10 millimeters (mm) square. But these general-use guidelines likely won't go far enough toward accommodating fingers made stiff with a startle response. It's old-fashioned tactile buttons that do best with users under a startle effect. It's one of the reasons that, despite the growing popularity of screen-based interfaces in cars, most vehicle interior design experts agree that the physical volume controls like that shown in Figure 2.8 are here to stay.

FIGURE 2.8
A physical volume dial is easier to manipulate than a touchscreen control while under a startle reflex.

"It's a safety issue," explained Florian Gulden, founder of Icon Incar, a Munich-based company that specializes in pushing forward car interior designs. "They need to be able to quickly adjust the volume if they are having a startle reflex." And that's because startle reflexes don't just affect motor skills, but also reduce a person's ability to use logical thinking. One study showed a startle reflex significantly reduced participants' ability to do math problems accurately for 17–30 seconds after the startle. In an avionics safety study, impaired judgment caused by bad startles was cited as a significant contributing factor in dozens of commercial airplane crashes.

Throughout these studies, it's shown that the length and severity of the impairment is directly related to the length and intensity of the startle. So, if a car volume is suddenly blaring at startling levels, you'll want to harness a user's lightning-fast reflexes to turn down the volume in milliseconds. This action gives drivers the best chance of limiting the severity of the startle and should improve their recovery time significantly.

Lessons from Analog

Although analog buttons and dials are demonstrably superior when designing for a startle effect, if you primarily design apps, websites, or software, it's unlikely within your purview to add or change physical buttons on the devices you design for. Fortunately, it's possible to translate some of the benefits of these well-studied analog interfaces to digital designs if you know *why* they work, as you'll see in Figures 2.9–2.16.

Beyond the physical design of the button, one of the best things an emergency stop button does is give users the comfort of knowing there's an escape route if things go really wrong. Providing a highly visual path to safety and support is often one of the best ways to make your users feel secure and increase trust in your designs.

FIGURE 2.9

The classic mushroom-shaped emergency stop button is purposefully oversized, and is made to be slapped with an open palm, not pressed with a single finger.

FIGURE 2.10

When designing a digital button for a critical fast action, use extremely short labels and extra-large buttons, the bigger the better. Leave room around the button if possible to accommodate stiff, clumsy fingers and overzealous hand movements.

FIGURE 2.11

Trip-wire style emergency stops supplement buttons on many factory floors because they can both be seen *and* reached by any worker within arm's length of the assembly line.

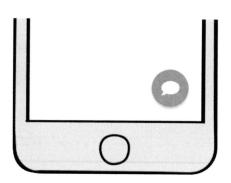

FIGURE 2.12

Consider ways to keep help within easy view and easy reach within your digital design. Something like the floating chat icon gives users an omnipresent escape route when they get lost or frustrated.

FIGURE 2.13

The bright red color of emergency stop buttons has been so well standardized and consistently regulated for the last 50 years that it has become intuitively associated with the idea of "stop" in all industrialized countries.

FIGURE 2.14

Beware of overuse of the color red in your digital interface as it can conflict with the intuitive knowledge that red means "stop." It's a common trap for brands where red is the primary brand color.

FIGURE 2.15

Emergency stop buttons are triggered with a press, but have a different mechanism for unlocking once the brake is engaged. To unlock the button above, it must be turned like a knob. This action protects against accidental brake release from "panic clicking" when someone who feels the machine is not responding fast enough may pound repeatedly on the button.

FIGURE 2.16

Consider points in your experience that may be trigger points for "panic clicking" or "impatience clicking." Code the system in a way that it won't turn off or restart the user's action on a repeat click. Whenever possible, give immediate visual feedback that the system is working and create a secondary path for canceling an action in progress if it was started in error.

PHOTO COURTESY OF JESSE RIESER

FIGURE 2.17

The pull-over button on the ceiling of the first Waymo taxi is easy to find, but also easy to forget.

Waymo, a Google company, was the first to bring self-driving taxis to consumers.

The "pull-over" button for Waymo's very first self-driving taxi is placed within easy view and easy reach of the back seat passengers (see Figure 2.17). High visibility of safety features like this are a big part of why this cutting-edge technology feels nonthreatening.

This button design is significantly more discrete than a factory floor override break, but its central placement is no accident. It's one of the first things that a passenger will scan for when entering a self-driving taxi for the first time. Once the location is noted, the low-profile, low-contrast design will help it fade into the background quite quickly. In contrast, a bright red button would be an eye-catching reminder that something might go wrong, making the rider feel they needed to stay alert. Waymo's goal is for you to relax and to stop thinking about the button altogether. The design and placement work to make the button easy to find, but also easy to forget.

However, forgetting the button and the dangers it brings to mind may make riders feel safer, but it won't make them actually *be* safer. As more driverless shuttles come along and consumers get more used to this modern miracle, they won't even think to locate the button until they are in a moment of need. If a low-contrast "forget me" design is combined with a different placement in every vehicle model, that will be incredibly dangerous for riders. It's unlikely they will be able to locate it quickly enough to prevent an imminent crash. Hopefully, regulators will soon establish standards that include a more intuitive red color, a standardized placement within easy reach of riders, and accessible safety accommodations for users with physical disabilities and vision impairments.

Protection from Jerks

Now that you've learned some of the science of the startle response, let's see if you can predict the outcome of this next experiment.

It was 1983, and scientists wanted to know if a mild startle could improve someone's ability to switch quickly between tasks. This particular experiment simulated the working environment of an air traffic controller at that time. Participants in the study were asked to perform one task at a desk in front of them, and then, when a signal sounded, turn in their chairs and start a second task at a desk to their left. One group of participants heard the signal at a reasonable 67dB (about the level of a bedside alarm clock). But for a second group, the signal noise blared out at a jarring 104dB, which reliably scared the crap out of participants.

Which group do you think switched tasks faster?

Surprisingly, the average time it took participants to switch tasks was nearly identical between the two groups. However, when the researchers looked more closely at their results, they noticed a very interesting difference in the data. Although the average was nearly identical, the spread in the switching time was different between the two groups. In the group who heard the nonstartling sound, the switching time was fairly consistent. But in the group who was startled, the difference in timing varied much more widely, as shown in Figure 2.18.

Task Transition Time

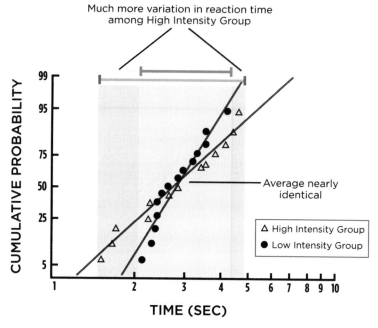

Much more variation in reaction time
among High Intensity Group

FIGURE 2.18

Task transition times varied widely between two groups in *Performance Recovery Following Startle*, a study by Richard I. Thackray, Ph.D.

Some participants were spurred to switch tasks much more quickly, and some participants were disoriented for a moment, which slowed their switching time. So, although a startle reflex improved the performance of some participants, it reduced the performance of others, essentially canceling out any benefits for the group as a whole.

Impulses May Vary

What task-switching experiments like this reveal is an important truth that designers must keep in mind—a startle reflex trades consistency for power. Think of it like changing your grip on a hammer: grab it close to the head, and you know you won't bang your thumb, but it will take many taps to get a nail in. Hold the hammer farther down the handle, and you can slam the nail home much faster, but increase the risk you will smash a finger. As a designer, it's important for you to understand just how unpredictable behavior can become in these circumstances.

THE RUNAWAY RENTAL

"We're going north on 125, and our accelerator is stuck," a strained male voice says on the 911 recording. "We're going 120," he tells the operator. "We're, we're in trouble, we can't, there's no brakes." The 911 operator asks the man if the driver can turn off the car, but before they can try that, the man says "We're approaching an intersection. We're approaching an intersection. Hold on." Then you hear yelling from the other passengers. He says, "Oh, oh, oh!" and then, slam. The call cuts off. All four passengers in that car died when it plowed full speed into another vehicle and then crashed into a ravine.

The 911 call recording from the Saylor family crash in the case above went viral in 2009 and what followed was a multibillion dollar scandal for Toyota known widely as the "Unintended Acceleration Investigation." Hundreds of people came forward claiming similar stories of runaway Toyotas and Lexuses of nearly every model. In total, ninety deaths were tied to unintended acceleration accidents in Toyota vehicles during the course of the investigation.

Toyota identified possible causes such as "sticky" accelerator pedals and sliding floor mats that may have explained how the cars kept accelerating even when the driver's foot was taken off the gas, and those mechanical issues were the basis of massive recalls and lawsuits. But while these mechanical issues explained the acceleration issues, they didn't fully explain why the cars weren't slowing down when the brakes were applied. Every car on the market is designed so that the brake will bring the car to a full stop, even if the gas pedal is fully depressed. After years of research, including a ten-month investigation involving NASA engineers, there remains no clear manufacturing or electrical system error that explains why the cars wouldn't just stop when the brakes were applied. Survivors of these crashes all claimed that they were pounding on the brake, but the car only went faster. So what was going on?

In his series *Revisionist History*, investigative journalist Malcolm Gladwell puts together a very compelling case that the root cause was simply human error. "Somewhere between intention and action, there's a garble, a glitch. And what happens? The driver puts his foot on the accelerator thinking it's the brake, he wants to stop the car, but, in fact, he's speeding it up." But how is it possible that so many experienced drivers might have made such a basic and deadly mistake?

It's actually a well-studied and very common phenomenon called *impulse variability*. Essentially, when you are performing a movement that involves multiple joints, like swinging a golf club, or throwing a ball, or slamming your foot really hard on a brake pedal, the larger the movement and the more intense your effort, the less accurate your movement will be. When a startle reflex superpowers your muscles, the effect can get even more exaggerated, causing actions like swerving toward danger or slamming the gas instead of the brake. It's not the brain sending the wrong instruction, it's the body executing the instruction poorly.

For digital designers, what this means is, the more animated the movements of your users, the more likely they are to click the wrong thing in your interface. As stress levels rise, and especially following any kind of startle or adrenaline rush, users need larger margins for error within your interface design. Accommodating this need most often translates to bigger buttons surrounded by increased white space.

DESIGN QUICK LOOK
TALES FROM USER TESTING: DEER DODGING

COURTESY OF BILL HOWARD

FIGURE 2.19
Ford's Virttex testing dome allows researchers to run simulations using realistic, prototype vehicles.

Ever wonder how life-and-death safety features are user-tested without risking people's lives? At Ford, they have a dozen high-tech driving simulators, including the testing dome pictured in

Figure 2.19 called *Virttex*. Full-sized car prototypes can be loaded into the dome with screens completely surrounding the vehicle. Hydraulics give test drivers a more realistic driving experience simulating the feeling of acceleration, deceleration, and turning into a sharp curve. When studying startle responses, researchers often attach biometric scanning sensors to the test drivers to monitor alertness and attention. Drivers may be asked to drive for eight hours or longer. Then, just as the biometrics indicate a driver is in danger of nodding off, the researchers use the simulators to have a deer jump in front of the sleepy driver, carefully observing their reflexes and reactions.

During his time as Global Director of HMI, Interactions & Ergonomics at Ford, Parrish Hanna saw all kinds of alarming driver behavior while observing closed course tests. He's seen test drivers "turn the wrong way, accelerate instead of breaking, and turn into a wall instead of avoiding it." Parrish believes these illogical behaviors are caused by a combination of factors. "I remember the wall on a particular turn that seemed to be magnetized to attract our test vehicle. There was something about the combination of requested tasks, plus the driving maneuver, plus the driver's state that turned the wall into a powerful magnet." Human behavior is affected by too many factors to ever be 100% predictable, which is why it's so important that vehicle safety systems are designed with multiple layers of protection that account for everything up to and including what may appear as self-sabotage on the part of the user, but is more likely an extreme case of impulse variability.

To Err Is Human, to Protect Design

In addition to the unpredictable physical effects, there is also evidence that a startle can make people extremely error prone in the moments directly following the startle. It is a highly studied phenomenon in the aerospace industry. At least five commercial airline crashes between 2004 and 2010 were deemed to be caused by errors (usually a series of errors) that the pilot made after experiencing a severe startle.

If you are an experienced UX professional, you've probably heard the saying, "There's no such thing as user error." But the saying doesn't mean that your users never make mistakes; it simply acknowledges that it's your job as a designer to minimize opportunities for missteps, help users quickly recover when they do get off track, and provide multiple layers of protection to ensure that your users never suffer catastrophic consequences from simply being a predictably imperfect human.

James Reason, one of the most well-known researchers of human error, organizes all the different types of mistakes that humans make into a very simple set of categories. The first category is *planning failures*. These are errors that happen because someone makes a bad plan based on faulty logic (*rule-based mistakes*) or false confidence (*knowledge-based mistakes*). Other times, a great plan is made, but it still fails in the execution phase. These *execution failures* happen either because someone has some kind of memory *lapse*, or they make an unintentional *slip*. These categories of human errors are diagramed in Figure 2.20.

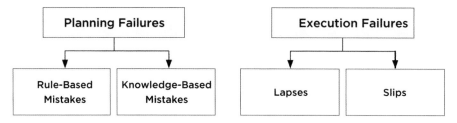

FIGURE 2.20

In his book *Human Error*, James Reason creates a taxonomy to better understand the types of possible errors and their root causes.

Here are examples of each of the four types of errors:

- **Rule-based mistakes:** Rule-based mistakes come from either the application of bad rules, such as thinking 2+2=5, or the misapplication of a good rule, such as thinking that all pharmacies are open 24 hours a day (because that is true in your city), only to pull into the parking lot of a small-town drug store late at night to discover it closed at 9 p.m.

- **Knowledge-based mistakes:** These are mistakes that happen when a situation requires critical thinking but the person uses mental shortcuts instead, such as falling for a trick question posed as a binary choice, "Which is heavier: a ton of bricks or a ton of feathers?" The answer, of course, is "Neither" since they both weigh a ton. (We'll talk more about mental shortcuts and bias in Chapter 3, "Intuitive Assessment.")

- **Lapses:** Lapses include all failures of memory, such as forgetting to put the cap back on the gas tank.
- **Slips:** Slips cover all types of unintended actions, including inaccuracy in a movement and doing one thing when you meant to do another. These errors are especially common when someone is trying to change a habit, such as accidentally driving to their previous office building the week after starting a job at a new company.

It's important to be able to distinguish between the types of errors because different errors require different interventions. Rule-based mistakes can often be circumvented through better education, while knowledge-based mistakes can be prevented through the introduction of processes that guard against bias. (See Chapter 5, "Reasoned Reaction," for more on techniques to reduce bias and encourage critical thinking.) Memory lapses can often be addressed through reminders. Slips, in contrast, while the easiest to catch after they've been made, can be the hardest to prevent prior to occurring.

Preventing slips can be a real challenge for designers because they are always unintentional, and they often happen in a split second. When heavy machinery is involved, slips can have dreadful consequences unless extensive measures are taken.

DESIGN QUICK LOOK
CUTTING OFF ERRORS

FIGURE 2.21
A SawStop is a powerful tool for contingency design in a woodshop.

SawStop, shown in Figure 2.21, is an excellent example of a design that can protect a user from grievous injury due to a slip. The system detects the instant the blade comes in contact with human skin and will stop a saw blade in less than 5 milliseconds. It completely destroys the blade in the process, but 10 out of 10 woodworkers agree that's a preferable outcome to sawing off their finger.

Even remarkable split-second interventions like the SawStop don't prevent the slip from happening; they only prevent the consequences of the slip from being catastrophic. Interventions like this fall under the umbrella of *contingency design*, helping users recover after things go wrong. Good contingency design will allow users to correct mistakes easily or warn users if they are about to make a choice that they can't easily go back on.

There is one piece of good news for digital designers who need to protect their users from the consequences of slips. While analog interfaces may be the best choice for capturing the benefit of the super speed that comes from a startle response, digital interfaces have their own unique strengths when it comes to contingency design. First, most mistakes made on a computer can often be easily reversed with an "undo." And for more permanent actions, "Are you sure?" pop-ups are fairly easy to add as an extra layer of protection. However, as research from the Nielsen Norman Group confirms, these types of pop-ups should be used sparingly, only for "actions with serious consequences," as overuse can build bad habits among users of clicking "yes" unconsciously, without any pause to read the prompt or consider the consequences. The real goal of these sorts of interventions is just to slow users down enough during an error-prone moment to give their rational mind a few extra seconds to catch up with their instinctive actions. This extra time allows the user to course correct on their own. (This technique of purposefully slowing a user down is often referred to as *adding friction* to a design.) If all else fails, digital features like "trash bins" and "drafts folders" are a great final safety net to allow users to recover accidentally deleted or dismissed work, even days or months after the fact.

Critical Information: Startle Reflex

When a sudden threat is perceived, a startle reflex is sometimes triggered, causing the body to move suddenly to protect itself. This subconscious reflex is triggered by a signal sent directly from the sensory thalamus to the amygdala, allowing the person to react faster than conscious thought. While the reflex is helpful in responding to the type of straightforward physical dangers that threatened our ancient ancestors, it is often less beneficial when dealing with the complex dangers of modern life.

What makes something startling and how can designers prevent or control a startle in their users?

Novelty, intensity, rise time, and priming are the critical factors in controlling a startle. The best way to make something less startling is to increase the rise time, making the change happen more slowly.

What are the pros and cons of a startle reflex and how can designers harness the beneficial aspects?

A startle has the beneficial effect of increasing the response time and power of gross-motor movements related to the fight-or-flight response, like running and punching. It has been shown to decrease the accuracy of fine-motor movements and impair cognitive function. Oversized physical buttons, like the iconic emergency stop button, are designed to take advantage of the gross-motor benefits of a startle. If physical buttons are unavailable, digital buttons should aim to be similarly enlarged, intuitively labeled, and within easy reach at all points in the experience.

How can designers protect users from the downsides of a startle?

When users are startled, you should anticipate that they will have a high rate of variability in their reactions and be more error prone, both physically and mentally. You can employ mitigation strategies appropriate to the type of error experienced to best protect users in those moments: rule-based errors through education, knowledge-based errors through processes that encourage critical thinking, lapses through reminders, and slips through contingency design such as SawStop, undo, or "Are you sure?" pop-ups.

Go to the Source

"Performance Recovery Following Startle: A Laboratory Approach to the Study of Behavioral Response to Sudden Aircraft Emergencies": A comprehensive deep dive into the startle response by Richard I. Thackray, 1988.

"Differential Effects of Startle on Reaction Time for Finger and Arm Movements": A study published by the *American Psychological Society* by Anthony N. Carlsen et al., 2009.

"The Effects of Startle on Pilots During Critical Events: A Case Study Analysis": A study by Wayne Martin, Patrick Murray, and Paul Bates, 2012.

"The Unexpected Physiology of Jump Scares": An Inverse.com article by Ben Guarino, 2015.

"'Go' Signal Intensity Influences the Sprint Start": A study on starter pistols and the startle reflex by Alexander Brown, Zoltan Kenwell, Brian Maraj, and David Collins of the University of Albert, 2008.

Revisionist History: "Blame Game": A podcast by Malcolm Gladwell about the Toyota Unintended Acceleration scandal, 2017.

"Inside Virttex, Ford's Amazing Driver Distraction Simulator": An article from *Extreme Tech* by Bill Howard, 2012.

Human Error: A book about all the ways that humans make mistakes by James Reason, 1991.

"Confirmation Dialogs Can Prevent User Errors—If Not Overused": An article from the Nielsen Norman Group, 2018.

CHAPTER 3

Intuitive
Assessment

As much as we humans like to think of ourselves as rational creatures, in control of our own minds, so much of what flits in and out of our heads comes from a place we have little conscious control over. When we get in a stressful situation or feel threatened, our brains often prioritize speed in decision-making, relying on the nearly instantaneous assessments that come from our subconscious, intuitive minds to guide our reactions to outside threats.

Helping users make decisions is the central goal for a wide range of UX designs—from ecommerce sites to medical devices. To guide users' choices effectively, it's critical to understand the factors that influence those decisions, both consciously and unconsciously.

The Science of Intuition

In the 1970s and 1980s, psychologist Gary Klein was one of the first in the modern era of Western medicine to bring human intuition, specifically the development of expert intuition, forward as a respectable topic of scientific study. He wanted to understand how experts made life-and-death decisions at critical moments. He and his associates interviewed hundreds of professionals in high-stress industries and what they found flew in the face of the accepted wisdom of how good decisions are made.

THE CASE OF...

THE PSYCHIC FIREMAN

Gary Klein didn't originally want to interview the firefighter about his extrasensory perception (ESP). Klein was trying to do serious science on the roots of expert intuition by interviewing fireground commanders about the tough decisions they made while on the job. A guy who thought he had psychic powers didn't exactly scream legitimacy. But this firefighter really wanted to tell this story as part of his interview, so at the end of the session, Klein finally relented.

This firefighter, he was unnamed in the study but let's call him Mike, had entered a home with his team to fight a small kitchen fire. They were spraying the flames with a hose, but the fire was not going out. Mike's ESP whispered to him that something was very wrong. He turned to his team and yelled, "Out! Everyone get out now!" They barely escaped before the floor of the kitchen collapsed straight down into the basement below.

By meticulously gathering, documenting, and dissecting hundreds of stories like this from professionals in high-stake fields, such as fireground commanders, tank platoon leaders, and nurses, Klein and his associates brought intuition out of the realm of the mystics and into the realm of science.

Their revolutionary insight was that experts in these high-stress fields rarely weigh multiple options when making critical decisions. They simply go with the first thing that pops into their mind that they believe will be successful, a process that would eventually be labeled *naturalistic decision-making*.

This method of decision-making is lightning fast, shockingly effective, and almost entirely subconscious. When Klein asked questions like, "How did you make the decision to enter the building from the back?" they would often respond with something like, "I didn't make a decision, I just did it." The interviewees had no memory of making a conscious choice, even though, when pressed, they admitted there were dozens of other options they could have used like entering via the front door, a window, a fire exit, etc. However, what mattered most in that moment was not a thorough consideration of all options, but fast action. When lives were on the line, these professionals spent little time worrying about finding the *best* option because they just needed an option—any option—that would work.

It turns out the source of the fire was not the kitchen. The fire had started in the basement just under the kitchen. What appeared to be a fairly small fire on the main floor was just the tip of a very fiery iceberg. Mike, of course, had no way of knowing this at the time—he didn't even realize the building had a basement. But as Klein pressed him for details during the interview, Mike kept remembering things about the fire that seemed off: the fire was extremely hot, much hotter than a fire of its size should be; it wasn't responding to the spray of water as expected; and, perhaps most puzzling, the fire was strangely quiet.

After the interview, Klein surmised that there were no psychic powers involved in this near-death escape. The extra heat, the unresponsive behavior of the fire, and the strange quiet of the flames, they were all clues. Mike may not have had enough time to consciously piece it all together, but subconsciously, his intuition recognized a deadly combination of factors, saving Mike's life and the lives of his teammates.

These insights directly contradicted the prevailing decisioning theory at the time, *comparative decision-making*, which stated that good decisions came from experts who listed all available options and carefully weighed pros and cons. However, this method is more time consuming and requires significant focus, creative thinking, and working memory. Those are three things that stress has a proven negative effect on. Through decades of study, Klein found little evidence to suggest that stress negatively affects the quality of *unconscious* choices made through naturalistic decision-making. So in times of high stress, in certain types of environments, intuitive decision-making may actually be *more* reliable than rational decision-making methods.

Klein's insights about the basis of intuition don't just apply to firefighters. In fact, most people go about their daily lives constantly making choices that don't look like choices. A rule of thumb used among researchers in this area is that about 90% of a neurotypical person's daily decisions are made entirely unconsciously through naturalistic decision-making. These individual decisions flow so effortlessly from one to the next that entire sequences can be completed with very little conscious intervention. Klein gave an example that many drivers have experienced—arriving at a destination with almost no memory of the trip there. Some people refer to this as being on "automatic pilot," and if you consider the level of complexity of tasks you can accomplish in this mode, the true power of intuition starts to become clear. Some of the smartest engineers in the world have been working to program computers to allow vehicles to drive themselves for decades now, and the machines can still only barely match human performance in extremely controlled environments. The human mind's ability to automate and layer activities is astounding.

Understanding intuition is key to being able to influence and leverage this human superpower. When it comes to intuition, the most important questions for designers are:

- How does intuition form?
- How do designers know when intuition can be trusted?
- How can designers help users develop reliable intuition?
- How can products support users in harnessing their intuition?
- When does user intuition become a harmful bias that can put users or others at risk?

Let's explore each of these questions more closely.

Pattern's Role in the Formation of Intuition

Intuition is built through lived experiences. From birth, your survival mind has been logging everything in your environment: categorizing objects and people, cataloguing threats and assets, and linking behaviors to patterns of cause and effect. When you enter a new situation, you instantly start the subconscious process of comparing what you see in front of you to patterns you've observed in the past. When you get a match, you apply whatever technique you used to survive the last time as a model to guide your behavior.

Whenever a user interacts with any kind of interface, physical or digital, they are adding to their mental library of interface patterns. Over the course of their lifetime, that library gets larger and larger. But unlike a physical library that takes more time to search the more books it contains, a mental library actually works *faster* as it grows. That's because the human brain arranges information via associations and similarities. So the more similar the interface is to a known pattern, the faster the subconscious makes the match. When a perfect or near-perfect match exists in the mental library, the understanding of how to use the interface springs instantaneously to mind with no conscious mental effort, and no uncertainty. As Klein puts it, "The accumulation of experience does not weigh people down, it lightens them up." When your design interface matches an interface that people have used before, users will start to call your design *intuitive*, even though it is their own carefully accumulated intuition that does all the work.

INHERENT MEANING FROM METAPHOR

COURTESY OF IVAN RADIC

FIGURE 3.1
Many electric vehicles are charged via plugs found in the same place that the fuel tank sits on gas-powered vehicles.

Metaphor can be a powerful aid when designing an intuitive interface for an unfamiliar product. Even though electric vehicles could
be charged from plugs located on nearly any part of the vehicle,
putting the charging port in the same spot that the fuel tank has
been on gas-powered cars leverages decades of muscle memory
and helps the new system feel immediately intuitive. In Figure 3.1,
you can see that even the design of the charging station mimics a
gas pump, down to the shape of the cord handle. Just as *skeuo-
morphic design* (digital interfaces that mimic the look of the physical objects they are replacing) fell out of favor as smartphones
were adopted by the masses, you can expect to see such a literal
transference of the fueling metaphor fade over time as well. In the
meantime, it's a great shortcut to usability for early adopters.

To leverage your user's intuition in your own designs requires an
understanding of what your user has and hasn't seen before. It's one
reason that designing for users from a very different culture or country than your own can be difficult. When you don't share the same
mental pattern library with your users, you can't count on your own
intuitive feelings about a particular design to guide your decisions.

In her talk "This Design Sucks," Netherland-based product designer
Camille Gribbons shared a cautionary tale of trying to design a hotel
booking site for an Indonesian audience from 11,357 kilometers away.
For example, their first design was a huge failure with Indonesian
users. The team was hamstrung by a poorly researched brief full of
false assumptions about the needs of the local audience and a project
manager that was not passing on feedback from the Indonesian
design team for fear that the scope of the project would expand and it
would take too long to launch. The Netherland design team's mini-
malistic Western aesthetic looked stark and untrustworthy to their
Indonesian audience who were used to websites full of bright color
and energetic animations.

The design was so different from what an Indonesian consumer
expected that nearly every user who looked at the original design
in usability testing done by the local team assumed it was a scam.
(This is a common intuitive response when user expectations are
defied. Assuming unexpected things are dangerous is what kept
our ancestors alive.) It wasn't until Camille flew to Indonesia, heard
the feedback from the local team first-hand, and took steps to fully

integrate the Indonesian design team into the design process that they were able to find success. The stark difference in style can be seen in the before-and-after pictures shown in Figure 3.2.

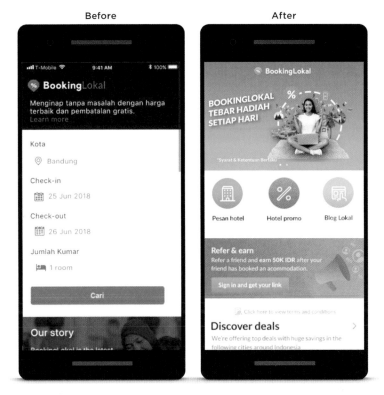

FIGURE 3.2
BookingLokal designs from before and after the integration of the Indonesian design team.

As important as empathy is in design, humility is just as critical. Be self-aware enough to know when your own intuition can be trusted and when it's more prudent to trust someone else's. Even immersive research techniques can't build a mental pattern library equivalent to a lifetime of living in a specific culture. As Camille learned, "Good design is definitely not the same thing everywhere in the world." That's why it's critical to involve people from the communities you are designing for early and often. This lesson also applies if you are building tools to be used by professionals doing a job you've never done before or consumers with a passion or background you do not share. Although it's important for designers to research and understand, it's equally critical to listen and accept recommendations from those with a background you lack. There simply is no substitute for lived experience.

When to Trust the Gut

Around the same time that psychologist Gary Klein was becoming a titan in the study of naturalistic decision-making, economist Daniel Kahneman was emerging as the definitive expert on irrational biases lurking in intuitive decision-making, work that would eventually win him the Nobel Prize.

What Klein called *intuition*, Kahneman called *bias*. Instead of highlighting life-saving moments of heroism in his work, he often delighted in studies where so-called experts, such as investment managers, were proven to be no better at predicting future outcomes than the flip of a coin. His research helped scientists better understand a myriad of cognitive biases, including confirmation bias (our tendency to trust data that confirm what we already believe and dismiss information that doesn't support our current position), availability bias (believing that if we can think of a lot of examples of something, it must be common, and if we can't think of any examples quickly, it must be uncommon), the anchoring effect (the tendency for the first piece of information offered to affect later judgments), and many others.

With such different outlooks on intuition, you might think Kahneman and Klein would give each other a wide birth, but they're actually more like academic frenemies. In the early 2000s, they decided to work together on an "adversarial collaboration" to cowrite an article exploring under what conditions (if any) an expert's intuition could be trusted. After enduring what Kahneman described as "long hours of discussion, endless exchanges of drafts, and hundreds of e-mails negotiating over words, and more than once almost giving up," the two eventually published the charmingly titled article "Conditions for Intuitive Expertise: A Failure to Disagree."

Through their work, Kahneman and Klein came to the agreement that it is possible to build expert intuition under certain conditions. You just need two things:

- "An environment that is sufficiently regular to be predictable."
- "An opportunity to learn these regularities through prolonged practice."

If even a world-famous skeptic like Kahneman is convinced that expert intuition is possible to develop under the right circumstances, then that is a compelling human superpower that designers are going to want to leverage if possible. But how do you know if you are designing for users in "an environment that is sufficiently regular to be predictable"?

To answer this question, it's imperative to understand that digital devices and apps are themselves objects that behave in extremely predictable patterns since they are programmed. This fact means that it's actually quite easy to build reliable intuition using digital interfaces with practice. (The only thing that would disrupt this process would be if the interface were buggy or deliberately coded to behave in a random fashion.) As of 2021, roughly half the world population has been using the internet for three or more years, according to the International Telecommunication Union. The majority of these users would qualify as having enough prolonged practice on computers and smartphones to have developed reliable intuition to apply across a wide range of screen-based applications. This knowledge allows them to predict pretty accurately how an app will work, even if they've never seen it before. The upside for designers is this means that you don't have to start at square one teaching someone how to use your product or developing that intuitive sense. The downside is that you're tethered to the conventions established by other designers, and if you stray too far from those "best" practices, you risk triggering suspicion, confusion, and even anger in your users. Even so, the answer is pretty clear-cut when it comes to digital designs—it *is* possible, even easy, for users to develop reliable intuition about digital interfaces.

Now, if you work on a product where the interaction is limited to a single user and a digital interface, then the discussion around your user's ability to develop reliable intuition can stop here. But many digital products are intended to support a user as they interact with external objects or other people—for example, air traffic control software, social media sites, intelligence analyst software, medical devices, or vehicle digital interfaces. If you work on a product like that, then there are two levels of intuition you have to consider:

- The intuition that guides the user's interaction with the interface.
- The intuition that guides the user's interaction with outside objects or people.

We've established that developing reliable intuition around digital interfaces is definitely possible. But how do you know when your users can trust their intuition about those outside elements?

The Kahneman/Klein study found evidence that there are certain fields in which a reliance on intuition can often lead to false confidence, and those fields include "stockbrokers, clinical psychologists, psychiatrists, college admissions officers, court judges, personnel selectors, and intelligence analysts." These jobs all share a need to try to predict future human behavior within an extremely complex web of influences, e.g., Will this student do well in college? Will this criminal reoffend? In many of these fields, it's still possible to make predictions that are better than random chance, but only if careful statistical analysis is applied and misleading attributes are ignored. If you design experiences where users must attempt to predict human behavior in high complexity environments like these, you'll want to steer users away from intuitive decision-making. Instead, focus more on the approaches outlined in Chapter 5, "Reasoned Reaction," that keep users focused on a systematic approach to problems.

Now, in some fields, Kahneman and Klein agree, intuition *can* be trusted to guide experienced professionals toward good choices. These professions include "livestock judges, astronomers, test pilots, soil judges, chess masters, physicists, mathematicians, accountants, grain inspectors, photo interpreters, and insurance analysts." The trends here are that these people work with things that currently exist (e.g., stars, planes, soil, cows), and there is a verifiable set of rules that are consistent, even if they are complex (e.g., chess moves, physics, math). When designing experiences for users, if the outside elements your user is engaging with are either items that behave in predictable patterns or people who are constrained to a limited set of choices (like in a game), then you should be able to help your users develop reliable intuition. Be sure to take full advantage of the intuition-building techniques discussed in this chapter, not just for creating intuitive interface designs, but for helping your users develop and utilize intuition about those outside elements as well.

Developing Intuition in Your Users

As a designer, you want to expand your users' intuitive knowledge because once something becomes intuitive, it frees up mental capacity that the user can apply elsewhere. Once a fighter pilot can intuitively fly a plane, they can concentrate on avoiding enemy fire. Once an online shopper can use an ecommerce site intuitively, it gives them more brain space to put together the perfect outfit, complete with accessories. No matter who your users are, the techniques for building intuitive knowledge are the same for any human and include the following essential elements:

- Repetition plus variety
- Clear, immediate feedback
- Stories and improvisation
- Interactive learning

THE CASE OF ...

THE TRANSFORMING RIFLE

Alexandra Gaski and her team were out on foot during a routine mission in Afghanistan. As the Intelligence Collector Specialist, she was assigned to comms and received the alert that there were Taliban members headed toward their location. "I let my commander know, and we started hightailing it back up the hill to the base."

Suddenly, she spotted a man carrying what looked like an AK-47. He ducked down behind one of the low mud walls lining a nearby field. "I raised my gun, and I started yelling, 'Gun! Gun! Enemy sighted! Position: Right. 40 meters.' I am ready to shoot this guy."

She was sighting down her rifle with her finger on the trigger when the man popped back up from behind the wall. Her instincts told her to hold for one second more. Heart pounding in her chest, she waited to see his muzzle swing down and point her way. But the man didn't shoulder his rifle.

"That's when I realized he was carrying a shovel, not a gun. He was a farmer."

Alex didn't shoot the man, but it was a close call. "My hand was there. My fingers were there. But I didn't pull the trigger."

When I asked her what stayed her hand, she gave credit to her training, running hours of shoot/don't shoot exercises. "Until I saw he had an intent to fire, I didn't want to pull the trigger. I'm really glad for the training and my innate sense of not wanting to kill another human being. But it easily could have gone another way."

Repetition Plus Variety

In order to build the instincts necessary to stay her hand in that critical moment, Alex spent hours upon hours training in the Army's digital simulators. Those simulators leveraged a combination of repetition and variety to train her and her fellow soldiers to distinguish civilians from enemies in the field. Holding a real weapon, modified to work with the computer, she ran through dozens of different simulation games where she had to distinguish noncombatants from disguised assailants, based on subtle patterns of behavior. Many times, the game would deliberately mislead the players. It might show what appeared to be a mother with her children, then the women turned, looking like she was asking for help, only to detonate her suicide vest. Or it might show a man walking erratically toward the soldier's truck despite the in-game leaders telling him to keep back. If the player held her fire long enough, she would discover the man was wearing headphones under his headscarf and had simply not heard the instructions to stop. When Alex was in basic training, she trained in front of a large projected screen. More modern military training programs can include virtual reality systems like that shown in Figure 3.3.

FIGURE 3.3
Virtual reality training equipment, such as these used by the U.S. Army, allows trainees to practice important critical-thinking and decision-making skills in simulated environments.

It was through simulations like these that Alex learned to look for not just one cue of aggression, but several together in different combinations, and she learned to recognize that some cues were more important than others. The esteemed economist Herbert Simon is often quoted as saying, "Intuition is nothing more and nothing less than recognition." Repetition with a liberal application of variety was critical for building a rich mental library of patterns that Alex could recognize. It's more than just the muscle memory of running the same scenario over and over again—the real value for making better intuitive choices came in the ability to run many similar variations and see the different outcomes. It's the reason that digital simulators are gaining popularity as training tools in a growing number of fields: flight, driving, medical training, and even customer service.

DESIGN QUICK LOOK
PRACTICING DIFFICULT CONVERSATIONS

FIGURE 3.4
Digital simulators like Safeguarding allow aid workers to practice critical communication skills around sensitive topics without the risk of retraumatizing survivors if mistakes are made.

UK-based BODYSWAPS develops VR simulator trainings that help people practice and improve interpersonal skills. Their program Safeguarding was created in conjunction with the Humanitarian Leadership Academy to help aid workers learn to navigate difficult conversations with survivors of exploitation and abuse. According to a case study released by BODYSWAPS, in the simulation, the trainee engages in conversations with virtual survivors of exploitation, giving the trainee a chance to practice having a sensitive conversation using their "own voice and body language." After the conversation is complete, the perspective of the trainee swaps places with the virtual survivor. They then "relive the conversation" from the perspective of the survivor, "seeing a replay of their avatar talking and moving the way they did." In addition to the opportunity for self-reflection and empathy building, the system provides feedback and recommendations for improvement. Since the training is digital and self-guided, trainees have the opportunity to run the simulation multiple times, testing different conversation paths and applying what they have learned.

Clear, Immediate Feedback

When you are creating interfaces meant to teach the difference between correct and incorrect choices, the faster and more clearly you can convey whether a choice is right or wrong, the faster the lesson will be learned by a user.

Duolingo, for instance, does a great job of giving clear and immediate feedback to users trying to learn a new language. Instead of waiting for the end of an exercise to get a total score, users get feedback question by question, as shown in Figure 3.5. This technique creates a near instantaneous feedback loop, significantly speeding up the learning process.

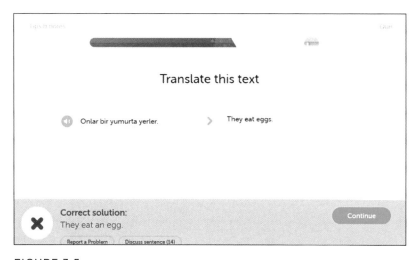

FIGURE 3.5
Duolingo gives learners immediate feedback after each answer in order to speed up learning.

Stories and Improvisation

Stories are powerful training tools that can teach concepts in a way that enables independent thinking and improvisation when things don't go as planned. It can be tempting when training a novice to just break down the information into a set of rules that you require the person to learn by rote. "When X condition is met, apply Y action." If you are creating training for positions with a high turnover, this actually may be the most effective way to train. However, if you are attempting to help a user really learn and internalize a new concept, this is not the best method for teaching intuitive understanding. Klein defines a true expert as someone who has the ability to improvise in a situation that goes outside of something they've seen done before. In order to improvise effectively, the person needs not only to know what the rule is, but also why it exists.

Stories convey basic rules, but they wrap the knowledge in context, so listeners learn the why, not just the what. Additionally, according to psychologist and author Jerome Bruner, stories are 22 times more memorable than facts alone, so they can significantly improve retention and recall.

THE CASE OF...

THE BLUE BABY

A NICU nurse sees a baby in her ward suddenly turn a terrible blue-black color. The baby's primary nurse believes it's a collapsed lung and calls for a doctor to treat it, but the nurse who first spotted the color change knows that it is pneumopericardium, which is air surrounding the heart, preventing it from beating properly. She'd had an infant patient die of this condition before, and she clearly remembered the look of the baby. Even though a collapsed lung is more common in an incubated patient, she had seen both situations before and knew exactly what was happening.

She tells her coworkers that it is the heart not the lungs, but they repeatedly point to the monitor that shows a steady 80 beats per minute. As the primary nurse yells out directions, the room fills with doctors and nurses springing into action to treat the child for a collapsed lung. The first nurse knows they are wrong, so she pushes forward with a stethoscope, screams for everyone to be quiet, and checks the heartbeat herself. She hears nothing and is able to convince the doctor to treat the child for pneumopericardium. "Stick the heart," she demands, slapping a syringe in his hand. He does so, releasing the air and saving the baby's life.

The "Case of the Blue Baby" is an abbreviated version of a teaching story that Klein shared in his book *Sources of Power* from his fellow expertise researcher Beth Crandall.

Klein's research team heard dozens of stories like this told and retold by hospital staff, passed down almost like fables. Although this story conveys the physical characteristics of a rare and hard-to-recognize condition, there are so many more lessons contained within this single story. Klein explains, "This story is a warning not to trust machines because they can mislead . . . It is a story about expertise," encouraging the listener to trust coworkers who have seen rare reactions over those who have not. "This is also a permission story. It tells when it is all right to make a fuss, to refuse to be reassured." By telling each other these kinds of stories, the hospital staff establishes cultural norms for their team. Storytelling is a rich, layered, and highly effective method of developing intuition that should be incorporated whenever possible.

DESIGN QUICK LOOK
STORYTELLING IN UI

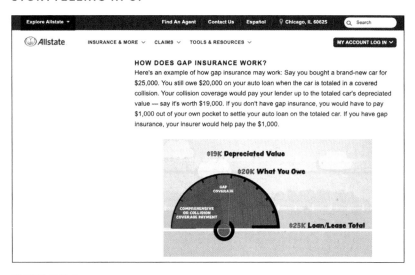

FIGURE 3.6
Allstate uses stories to explain how complex insurance coverages work.

Good UX writing is often thought of as short, simple, and to the point, but storytelling certainly has its place in interface design. In my work at Allstate, I found some of the most positive customer comments about written content came from user testing the little stories that explain how coverages worked in the case of an accident, like the one shown in Figure 3.6. They really helped make the complicated concepts of insurance more accessible, giving our customers a deeper understanding of what they were buying and how it could be used. Consider the use of example narratives in your own products. They perform well in FAQs and customer support forums.

Interactive Learning

The incorporation of interactive learning has significant advantages when attempting to help someone gain intuitive knowledge. When you learn something in a classroom setting, you primarily log that information through your prefrontal cortex. When you learn by doing, you also log that information in your hippocampus. The whole lightning-fast, subconscious, pattern-matching process we've been discussing in this chapter happens primarily in the hippocampus (see Figure 3.7). You could say it's the home of your intuition.

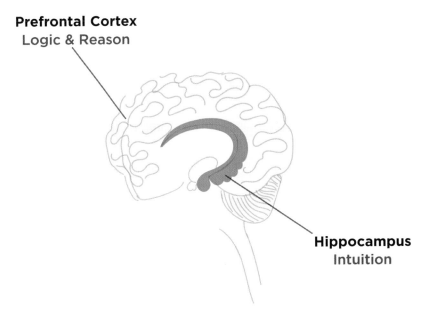

Prefrontal Cortex
Logic & Reason

Hippocampus
Intuition

FIGURE 3.7
The prefrontal cortex is located at your forehead, whereas the hippocampus is more centrally located, just above the spinal cord.

When a stress response is triggered, blood is pulled from your extremities toward your core. That's why your fingers get cold and clumsy when you're nervous. This pattern is also seen in the brain. Blood moves from the outer edges of your brain, like the logic centers in your frontal lobes, toward the mid-brain where your hippocampus sits. This cardiovascular shift makes it more difficult to think rationally, and it leads to more intuitive decision-making.

That's why it's so important that people who need to recall certain knowledge and exhibit certain behaviors in times of stress learn through drills and practice, not just listening passively to lectures. At the extreme end of the scale, you've got soldiers doing full-on tactical exercises so they can develop muscle memory that can be relied on in the middle of a war zone. In more everyday applications, a user who actually clicks and uses a UI element during a new customer tutorial will be more likely to be able to find that feature again a month later when they return to the site in a panic to pay a late bill. No matter what you're trying to teach someone, the more hands-on the lesson is, the more likely they will be able to recall that lesson when they are stressed.

Harnessing Intuition

As I draft this very sentence, my brain and body are functioning on multiple levels across a spectrum of conscious and unconscious control, as shown in Figure 3.8.

Simultaneous Actions: Conscious to Unconscious Control

FIGURE 3.8
Humans function through complex layers of automation, allowing the conscious mind to focus intensely on just the most critical and complicated tasks.

Humans have a very limited amount of working memory; however, they have a shockingly large capacity to learn and automate skills. Once they can do something fairly automatically, that frees up working memory that they can use to layer in additional actions and focus on higher level concepts. The less that people have to think about semi-autonomous actions, like typing and spelling, the more brainpower they have left to convey interesting insights and witty turns of phrase.

The real power of intuition starts when someone masters a skill to the point of *automaticity*, allowing them to perform tasks intentionally but with very little conscious effort. The knowledge or action they need springs effortlessly forward the instant they need it. This action frees up working memory for higher level concepts and increased situational awareness.

One of the worst things a design can do is to interrupt that automaticity—especially if you are designing a digital tool meant to be used by skilled professionals. Professionals in all types of fields work hard to master the skills of their trade. Automaticity is part of that process of mastery, and when it is achieved, it allows the professional to focus their attention on solving complex problems.

Erik Pedersen, a product design manager, recalls observing customer service reps on the job and the extreme familiarity they had with their software. "They knew the software so well they had memorized where the buttons were located from page to page. They would move their mouse to where they knew the button would be and the instant the page loaded, they would click it." This muscle memory allowed the reps to fly through the software, even while handling complex conversations on the phone with customers. Once Pedersen observed this behavior, he understood that he and his team would need to be circumspect when making iterative changes to the software—frequent updates that shifted locations of buttons along the critical path would destroy the proficiency these professionals had gained in their work environment and slow down their processes significantly.

Beyond interrupting muscle memory, creating interfaces that force someone to overthink the individual steps of their process can also cause major interruptions in their ability to perform that process. If you really want to ruin a pro athlete's game, ask them to explain their technique to you right before they go on the court or field. When an expert "gets in their head" about steps that would normally be performed in an automatic fashion, that steals mental capacity away from the really critical parts of their jobs.

Rhett Rapier, a usability specialist with Johnson & Johnson Medical Devices, has observed more than 100 surgeries in operating rooms (ORs) across Europe, and he's seen too many instances like the one in the following case study where a poor setup in the OR made procedures much harder than they need to be.

THE CASE OF...

THE BACKWARD SURGERY

In an operating room in southern Germany, a trauma surgeon was reassembling a shattered tibia. To finish the surgery, she needed to drill a hole into the bone and insert a screw to hold a metal pin in place. Because this was a minimally invasive surgery, her 5mm-wide drilling target was buried under cortical bone. She had to rely on static X-ray images to know if her drill bit was anywhere near its mark.

Unfortunately for this young surgeon, the X-ray display screen was not positioned well in the OR. It was nearly directly behind her. When she carefully placed the drill bit on the bone surface and then asked for an X-ray to check her positioning, she was forced to peer over her left shoulder to see the image that appeared on the screen. The drill bit was far from its desired location over the metal pin.

The surgeon made an adjustment left when she should have moved right.

Another X-ray.

Seeing her mistake she pushed her drill back toward the target, but it still was not quite right.

Another X-ray.

An overcorrection.

Another X-ray, and another, and another.

Each time she looked over her shoulder, the surgeon was leaning heavily on her spatial reasoning to mentally flip the image, to envision it from her body's current orientation. She had to insert four separate pins through the procedure, the mental effort was draining, and she could tell her accuracy was getting worse the longer the surgery dragged out. Her eyes flicked to the clock. A surgery that should have taken no more than 30 minutes was already passing the 60-minute mark. The surgeon took a deep breath and continued her work.

"Orientation and proximity are key to usability," Rhett explained, "If our devices don't help orient users naturally and function in close proximity, difficulties and even errors are sure to follow." Positions of screens are usually practical—close to a plug or a storage closet—but aren't always ideal for surgery.

J&J and other medical device manufacturers are beginning to change this, creating small screens (such as the one shown in Figure 3.9) that are specially made to be used within the *sterile field*, the area directly surrounding the operating table that is kept germ-free.

FIGURE 3.9
This operating table screen from Philips is designed to sit within the sterile field.

Additionally, sensor technology is used to coordinate devices and software to create a real-time visualization that orients the surgeons intuitively as they place implants inside the body. This solution gives surgeons a stream of visual feedback while avoiding the need to expose the staff and patient to excessive radiation from X-rays. When the surgeons, screens, and tools are in proper alignment and visual feedback is in real time, the surgeons can make adjustments completely intuitively and work much faster, significantly decreasing the likelihood of error.

Interfaces like these are a great example of what it looks like when a design is harnessing intuition. They take a task that was difficult, awkward, and time consuming and make it effortless. From the first time a human sharpened a rock to help them smash open a coconut, this is what interface design has been about—helping people do stuff more easily. And each time a design reduced the amount of effort it took to do one task, humans put that extra energy into advancing on to the next thing and the next until, before you knew it, we were sending our coconut-smashing selves up into space.

To truly harness intuition in your interface design is to create a design that is nearly invisible to the human using it—something that becomes an extension of themselves. Designing for intuition requires a certain amount of humility, because when it's done right, your design becomes something the user barely notices. The best compliment you can get from a user in usability testing is when they successfully complete a task and when you ask about the interface for that task they say, "Oh, I didn't even notice it." That's the mark that you are truly tapping into intuition. That's the mark of a successful design.

Consider the moments that your user is expending the most mental or physical effort. How can your design give users exactly what they need in that moment? How can you convert that effort into something effortless?

DESIGN QUICK LOOK

INTUITION DRIVING DESIGN

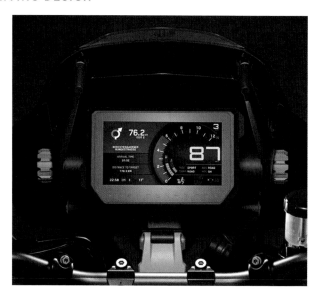

FIGURE 3.10
Some motorbike displays make the most of the small screen size by automatically adjusting the display based on the context and speed of the bike.

When designing a compact interface for a motorbike that will be racing along the autobahn, you've got to understand the specific needs of your rider and how those needs change at different moments throughout the drive. Interface designer George Cave explained, "As the speed of the ride increases, so does the rider's need to make decisions fast and intuitively." In the interface created by George's team at Austrian design firm KISKA, as the rider reaches high speeds, the display simplifies, removing superfluous icons, and it increases the size of the one thing the rider cares about in that moment—the speedometer reading.

Motorbike companies are also experimenting with enhanced safety features, like the ability to warn a rider if they are approaching a hard turn. But because speeding along curvy mountain roads is one of the favorite activities of their target user, the system can't be flashing a "Warning, hard turn ahead," constantly. Many riders actually choose routes that maximize the number of curves, and they take them at breakneck speed, seeing how low they can get their knees to the pavement as they lean into the turn. If the bike interrupted their focus on every curve, the joy would be lost, as would the automaticity needed for a smooth expert performance. Designers are finding ways to go the extra mile (or maybe I should say extra kilometer), dreaming up features like special maps of European highways that allow the system to know exactly how fast is too fast for each curve. That way the system only alerts the riders if they are approaching a curve at a truly dangerous speed.

When Intuition Cannot Be Trusted

While human intuition is fast and powerful, it is not infallible, far from it. Humans are particularly adept at finding patterns of cause and effect, but in an environment that has an unusually high amount of randomness, they have a tendency to find patterns that don't actually exist. That's when they drift into the realm of superstition.

In 1948, Burrhus F. Skinner ran a groundbreaking (but also hilarious) behavioral psychology experiment known as the *Superstitious Pigeon Experiment*. He put hungry pigeons in a cage and taught them that if they clicked a button, their food would be dispensed. Then he took away the button and switched the food to be dispensed automatically every 20 seconds. But these birds had learned that action equaled reward. They didn't realize the reward now happened automatically. Instead, if a bird happened to flap their wings just before the food was dispensed, then they associated that movement with getting the food. Wanting more food, they'd do it again, and eventually more food would drop. They quickly began to believe that repeating that action was causing the food to be delivered. The pigeons in the experiment began to display all sorts of odd, ritualistic behavior—spinning in circles, swinging their heads back and forth, bobbing up and down—and would repeat these behaviors incessantly, never making the connection that the treat came automatically, even if they did nothing at all.

Follow-up studies have shown that humans are just like those pigeons. They are similarly prone to draw false connections between their actions and outcomes in environments in which they actually have no control. The best way for you, as a designer, to prevent the development of superstition is to help your users clarify cause and effect.

Separating Superstition from Intuition

There's probably no industry that has its roots deeper in superstition than agriculture. For most of human history, the success or failure of a crop was attributed to virgin sacrifices, spells, prayers, dances, or the whims of the gods. While there were plenty of things humans could actually do to influence the success of their crops—watering, feeding, weeding, etc.—there was also so much randomness introduced from factors like the weather, biology, pests, soil content, and more that it took countless generations to sort out the true causation from mere correlation. Agricultural scientists continue making new discoveries even today.

Agritech software like Agworld, shown in Figure 3.11, now allows farmers to track input and output on their fields at a level of detail their forefathers could never have imagined. The industry refers to this kind of app as *precision farming software*. Of course, humans are as prone to find false patterns in big data as any other context, but the level of detail and the ability to track changes year after year really help farmers see both the literal and figurative fruits of their labors.

In an Agworld case study, Hunter Current of Alturas Ranches shared how the software had changed the way they did business. "When we used to plan and report, we just took the field average—which was the best data we had available. But, especially with alfalfa hay and using one stack yard for multiple fields for example, this quickly got very inaccurate." When data were generalized or inaccurate as Current described, it was very easy to draw false conclusions about cause and effect.

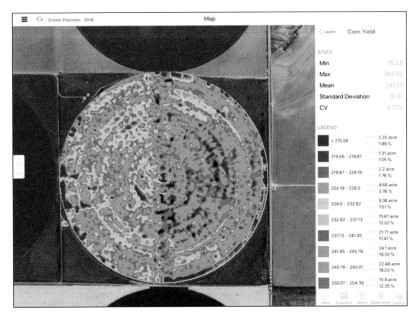

FIGURE 3.11

Agworld precision farming software visualizes crop production from every acre of the farmer's field.

Mark Monmonier, author of *How to Lie with Maps*, points out that data maps can *only* show correlation and associations. Causation, he says, "depends on logic and supporting evidence." To prove that one action actually caused a reaction takes critical thinking that must come on the part of the farmers. The precision software can help immensely in this endeavor by allowing farmers to zoom in on one aspect at a time. Using a wide variety of sensors and data streams, farmers can isolate all kinds of variables—from water to soil chemistry to weather to fertilizers—and see how they relate to detailed yield measurements.

"The way Agworld presents the information sparks inquisitiveness, rekindles memory, and naturally creates discussion among my crew," Current said. "When you can see a field that's dark green and the yield per acre is high, you visually imprint that on your mind to say yes—that's where our stronger soils are, that's where we get a better irrigation, or that's where we did a fertilizer study, for example."

Detailed visualization tools help simplify an immensely complex environment and help these farmers avoid confusing superstition with intuition.

Exposed Bias

Biases are mental shortcuts that happen largely on the subconscious level, which is a pretty accurate description of intuition as well. You could argue that bias and intuition are two words for the same mental processes, although obviously the two words have very different connotations for most people. That's because when people discuss bias, they are often focused on harmful biases. These take many forms, including the kinds of biases that lead to overconfidence and bad decisions, such as the confirmation and availability biases mentioned previously. But the type of harmful bias that tends to get the most attention in popular media is *stereotyping*, especially stereotyping people. Examples of harmful stereotypes include "mechanics are men" or "Asian people are good at math" (yes, even positive stereotypes can be harmful). If shared by enough people, these types of assumptions can change the course of someone's life or career by boxing them into limited roles and behaviors based on how they look or what groups they are a part of. On a larger scale, stereotypes like the ones that make false links between race and violence influence critical policies like immigration and policing. These types of false biases, if held long enough by enough people, can shape society and lead to institutional-ized structures of oppression that span across generations.

Biases, both harmful and benign, are formed through pattern recognition and categorization. This process of categorization results in a brain filled with stereotypes—many of which are actually quite helpful. For example, when you see a horse, you know it's a horse not because you've been told that particular creature in front of you is a horse, but because it has stereotypical characteristics of other animals you've encountered in the past that are called "horses"—four legs, a large body, etc. You likely also hold assumptions that can help you anticipate its behaviors—that it can run faster than you, that it will make whinnying noises, and so on.

For the most part, this use of a mental shortcut is helpful, allowing you to quickly understand the things that surround you and antici-pate how they will behave or how you can use them. The problem comes when people become overconfident in these assumptions and don't apply an appropriate amount of critical thinking. Stereotypes aren't guaranteed to be correct—perhaps the horse has an injury that would prevent it from running. Or maybe it's not a horse but a mule or zebra. Every stereotype-based thought that springs to mind needs to be considered critically in order to place the appropriate amount of confidence in the assumption being offered by your intuition.

As Klein and Kahneman concluded during their adversarial collaboration, these types of pattern-based assumptions (which Klein called *intuition* and Kahneman called *bias*) are fairly reliable when applied to predictable, uncomplicated things and become less reliable as complexity and variety are increased. Which is why it's so risky to apply stereotypes to humans—humans are both highly complex and extremely diverse.

There is no way to prevent a mind from offering up intuitive, stereotype-based thoughts (nothing short of brain damage at least). But as a designer you can influence the *formation* of these stereotypes by feeding new data into the mental pattern library of your user through education and training. The designers in this next story attempted to do just that.

THE CASE OF...

THE MISSING CHEST

Sarah Collinson vividly remembers the day the fake breasts arrived at her office. Sarah leads a design team at JOAN, a full-service, female-led creative agency in New York. When she walked into work that day, she found her coworkers standing around a large table in the center of the office that was covered with dozens of prosthetics made of plastic, foam, silicone, and rubber. One colleague pressed the heel of her hand down into a breast, testing the resistance. Two others discussed the durability and availability of foam vs. silicone breasts. Sarah was thrilled the project had finally reached this stage. She had high hopes for this table full of boobs—they were going to save people's lives.

According to a study by Dr. Audrey Blewer of Duke University, women are 27% less likely than men to receive CPR from a bystander. This fact is partly due to a common misconception that women don't have heart attacks (when, in fact, one in four women in the U.S. die from heart-related issues). A concern for hurting the woman was cited as another factor, but there was a third big issue holding bystanders back. Performing CPR on a person with breasts (not everyone with breasts identifies as a woman) requires directly touching their left breast. The correct hand position is shown in Figure 3.12. Now that might not be something someone would hesitate over if it came to saving the life of a person they know, but the study revealed that when it came to performing CPR on a stranger, a concern for being seen to touch that person inappropriately was actually holding a significant number of people back from performing this life-saving intervention.

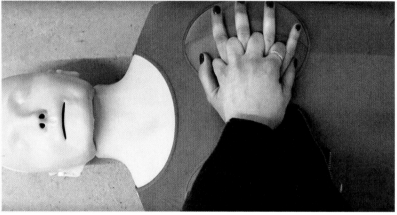

COURTESY OF JOAN CREATIVE

FIGURE 3.12
Proper CPR technique is modeled on a Womanikin.

"We needed to normalize giving CPR to a woman. To introduce breasts into that situation and touching breasts in a nonsexual way," Sarah told me. The majority of CPR manikins used in training classes are flat chested, so even if the topic is discussed during training, trainees almost never actually practice their technique on a full-figured figure. As discussed previously, if you want someone to learn information that sticks with them in a crisis, it's best if that information is learned through hands-on practice.

To enable this, the team at JOAN, working with the gender-equality group United State of Women, developed Womanikin, a low-cost wrap that could be easily added to a flat-chested CPR dummy (shown in Figure 3.13). This simple modification gave trainees a chance to learn proper technique, get over their discomfort, and be empowered to save the lives of more people with breasts.

COURTESY OF JOAN CREATIVE

FIGURE 3.13
Womanikin was designed by JOAN Creative in partnership with United State of Women.

They have posted their open-source, simple-to-sew instructions online for anyone to download, make, or even sell. The medical community is celebrating the project. Dr. Robert Glatter, an emergency physician at Lenox Hill Hospital in Manhattan, told MarketWatch, "The concept is excellent for two reasons: it will help increase adoption of CPR in women, but also aid in reducing gender bias regarding heart disease in women."

Beyond the ingenuity of the solution, there are some deeper lessons for designers here revealed by considering *who* was harmed by the bias in the original flat-chested manikin design. It was not the person who bought the dummy (the CPR trainer), nor was it the direct user (the student in the CPR class). The person harmed was the heart attack victim whom the student didn't try to save. As a designer, you may think about your users and customers all the time, but who are all the people who never use your designs but still feel their effects?

Truly empathetic designers measure their success in terms of the ultimate goals of their users. For example, people who take a CPR class don't do it just so they know CPR; rather, they do it so they can save a life. When you are searching for harmful bias within your own designs, it's important that you are measuring the right things. When you measure the *impact* of your designs, your metrics go beyond simple usage or task completion. You must measure what matters to the people who use your designs.

DESIGN QUICK LOOK
THE MOST ANNOYING SOUND IN THE WORLD

FIGURE 3.14
When chip-enabled payment devices began appearing in retail stores across the U.S., they brought with them what some have called "the most annoying sound in the world."

If you have made a purchase with a chip-enabled credit card in a U.S. store since the 2015 rollout of the chip-reader credit card machines shown in Figure 3.14, your ears have likely been assaulted by the annoying "eh, eh, eh, eh" sound that plays each time a transaction is completed. While this sound cue was originally designed with the end-user in mind, trying to prevent customers from leaving their cards in the machines, it was actually the cashiers who were most affected by this sound design choice, having to listen to that sound dozens or even hundreds of times a day. Having such a harsh noise constantly play throughout their shift significantly lowered the quality of the work environment for these employees. Given the high number of people whose work environments were affected by this and the relatively low number of customers who would leave behind their cards in the machines, this design decision seems to be biased toward prioritizing the short-term convenience of the customer rather than the long-term comfort of the cashiers.

Critical Information: Intuitive Assessment

Klein's research on intuition shows that 90% of the decisions most humans make happen automatically, instantly, and with no conscious effort. These are things that people "just know" they should do next. This subconscious knowledge is called *intuition*. People often rely more heavily on intuition in stressful situations that require fast decisions and fast action. Once they learn something intuitively, their mind can go on "autopilot" for that activity, which frees up their minds to focus on more challenging tasks. Humans have an astonishing ability to "stack" activities and handle increased complexity through this process.

There were several important questions about intuition posed at the top of the chapter. Let's revisit them.

How does intuition form?

As Simon says, "Intuition is nothing more and nothing less than recognition." The more patterns and experiences someone is exposed to around a specific topic, the stronger their intuition becomes in that area.

How do designers know when their users' intuition can be trusted?

Kahneman and Klein tell us that intuition can be developed in any environment "that is sufficiently regular to be predictable" and by individuals with "an opportunity to learn these regularities through prolonged practice." Users can definitely develop reliable intuition about how to use interfaces as long as the interfaces behave in a predictable pattern and users have enough exposure to the interfaces to build proficiency. Users can also develop reliable intuition about how to use digital tools when interacting with objects and people as long as those objects and people behave in predictable ways. When it comes to predicting future human behavior in complex circumstances with high variability, intuition is rarely reliable.

How can designers help users develop good intuition?

Intuition is built through lived experiences, not book learning. It is best developed through muscle memory, physical interactions, and actively engaging with an interface. Design techniques for developing intuitive knowledge in your users include repetition plus variation, clear and immediate feedback, stories, and interactive learning.

How can products support users in harnessing their intuition?

If you want your designs to harness the power of human intuition, you should focus on creating interfaces that are effortless and unobtrusive. Whenever possible, use familiar, common patterns. When you must create new UI patterns, utilize metaphors to link the new interaction to something already familiar.

When does intuition become a harmful bias that can put users or others at risk?

There are certain environments where relying on intuition can lead users to harmful bias or superstitious thinking. These kinds of mental traps often show up when users are trying to draw connections between cause and effect in high-complexity, high-variability situations or use stereotypes to predict human behavior. To address superstition, you should give users tools to separate cause from effect. To address harmful behavior caused by false assumptions and bias, expose users to hands-on experiences that retrain their brains away from the harmful biases they might hold.

Go to the Source

"Conditions for Intuitive Expertise: A Failure to Disagree": Research performed as an adversarial collaboration by Daniel Kahneman and Gary Klein.

Sources of Power: How People Make Decisions: A book on the science of intuition by Gary Klein.

"This Design Sucks"—YouTube: A talk by Camille Gribbons at Amuse Conference.

Gut Feelings: A book about intuition and behavior by Gerd Gigerenzer.

"'Superstition' in the Pigeon": A foundational study on how superstition is formed by B. F. Skinner.

Actual Minds, Possible Worlds: A book by Jerome Bruner that explores, among other topics, the power of stories to teach, 1987.

How to Lie with Maps: A book about data visualization and map design by Mark Monmonier.

"Gender Disparities Among Adult Recipients of Bystander Cardiopulmonary Resuscitation in the Public": The study by Dr. Audrey Blewer that inspired the Womanikin.

CHAPTER 4

Fight, Flight, or Freeze

"How was a 20-year-old with no income able to get assigned almost a million dollars' worth of leverage?" This pointed question was found within the suicide note of college student Alexander Kearns. Kearns downloaded the popular Robinhood stock-trading app to try his hand at investing while stuck at his parents' house in 2020 during the coronavirus lockdown. Robinhood used commission-free trades and a gamified interface design to make stock trading more accessible in order to appeal to Millennial and Gen Z users. However, they may have made it a bit too easy. With a few clicks, Kearns was able to unlock high-risk trading features he didn't fully understand, like options trading. Just a few months into his experimenting with the app, Kearns opened it to see that he had a cash balance of –$730,165.72, as shown in Figure 4.1. He interpreted this to mean he was now three-quarters of a million dollars in debt.

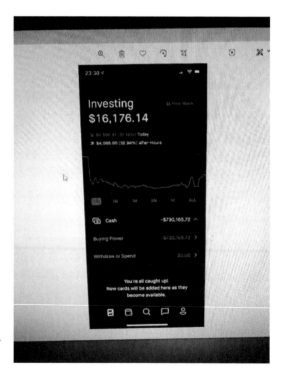

FIGURE 4.1
This screenshot from Kearns's account was shared by Bill Brewster on Twitter.

"When he saw that $730,000 number as a negative, he thought that he had blown up his entire future," said Bill Brewster, Kearns's cousin-in-law, explaining what drove the young man to suicide. "Tragically, I don't even think he made that big of a mistake." Although Robinhood has declined to publicly share details of Kearns's account, citing privacy concerns, experts speculate that the interface design of the Robinhood app didn't properly convey to Kearns what was happening with his account. According to an article from *Forbes*, "It may not have represented uncollateralized indebtedness at all, but rather his temporary balance until the stocks underlying his assigned options actually settled into his account." In other words, it's likely that if Kearns had waited for the trades to finish processing, his negative balance would have been significantly reduced, even wiped out completely.

Kearns's experience with the Robinhood app triggered a *fight-or-flight response*. The lawsuit filed by his family against Robinhood refers to his increasing levels of "panic" seven times, and *panic* is defined scientifically as an extreme fight-or-flight response leading to drastic, irrational behavior. Many people think of fight or flight as being triggered by outside physical threats, like a tiger attack, but threats to your personal resources such as money, food, or other rationed goods also can trigger a response, as can threats to your social standing. Additionally, fight or flight can be caused by internal feelings of guilt, shame, embarrassment, despair, and anxiety.

Whether the response is externally or internally triggered, the primary purpose of the response is to help the person experiencing it escape or subdue a perceived threat, even threats caused by their own emotions. This is why people sometimes run out of the room when they become embarrassed or avoid doing important but emotionally difficult tasks—that's the flight response at work. Similarly, some people experiencing strong negative emotions become self-destructive, and suicide is at the extreme end of this internally directed fight response. As explained in an article by Sanchez Federico in the *Annals of Behavioral Science*, "Uncontrolled aggression is channeled against the source of the threat...which is being generated by one's own brain. Suddenly, extinguishing the root of this threat becomes absolutely imperative. An attack against oneself, suicide, is the focus of the aggressive behavior." Federico goes on to clarify, "It is easy to try to see this as a form of self-hate, but aggression against a threat (even an internal threat) is more in line with this behavior."

The human behaviors triggered by the fight-or-flight response are often difficult to predict and vary widely, not just between individuals, but from situation to situation. While designing for users in this state may be challenging, fight or flight is far too prevalent and important a factor to ignore or avoid. Its influence on user behavior can be seen in nearly every stressful interaction. In this chapter, we'll address these critical questions about designing for a user in the midst of a fight-or-flight response:

- How does the fight-or-flight response work and what is a designer's role in managing its effects?
- How should designers handle users who are trying to *flee* from a product or experience?
- What de-escalation techniques can calm a user in *fight* mode?
- How can designers help a *frozen* user get unstuck and take positive action to address their stress?
- What are specific UI design techniques that make interfaces easier to use when someone is in a fight-or-flight mindset?

How the Fight-or-Flight Response Works

When humans find themselves in a situation they perceive as threatening, their fight-or-flight response causes them to respond in one of three ways. First, they try to flee. If that proves impossible or if fleeing has too many risks, then they turn to fight. There is a third option as well. Occasionally, their system is so overwhelmed by the scale of the threat that neither fighting nor fleeing seems a viable option. In these cases, they shut down. This is sometimes referred to as a *freeze response* or *tonic immobility*, similar to a deer in the headlights. These impulses always come in the same order: 1. Flight, 2. Fight, 3. Freeze. (Though many modern scientists and academics have switched to the more accurate label of "flight-fight-freeze response," in this book, I'll stick with the colloquial "fight-or-flight response.")

Most behaviors that users exhibit in response to stress link back to one of these three responses. Small amounts of stress manifest in small ways, like leaving a frustrating website (flight), posting rude online comments (fight), or being so overwhelmed by the number of choices for a product that nothing is purchased at all (freeze). As stress increases to the level of a true life-and-death threat, the reactions become much more literal: running away, throwing punches, or entering a catatonic state.

It is extremely difficult to predict which of the three reactions will manifest in any given person or situation. Since the "choice" happens instantly and entirely subconsciously, it can be difficult to study.

There is still much to be learned about how it works. Scientists do know that mammals, including humans, will instinctually flee danger unless escape seems impossible or more dangerous than staying and fighting. They also know that the fight response will override a flee response in cases where the mammal is protecting an individual or group it is bonded to, especially offspring, family, and fellow fighters. And finally, they know that the freeze reaction is relatively rare compared to either flight or fight. While some animals like opossums use tonic immobility as a go-to defense mechanism, in humans, it's typically a last resort. Its main function is to help a victim minimize physical and mental trauma suffered during attacks that are both unwinnable and inescapable.

A Designer's Role

It's rare for users to be in a full fight-or-flight response while using a digital interface. However, many users turn to their smartphones in the immediate aftermath of an accident or crisis. They may be calling for emergency services, but they might also be contacting loved ones for help through a variety of methods, or documenting—even live-streaming—the scene as a situation unfolds. And then there are cases like the Robinhood example where the fight-or-flight response is caused by the app itself delivering shocking or terrible news.

Because of the extreme, unpredictable, and often destructive behaviors that humans exhibit while in a fight-or-flight response, a designer's primary goal regarding the response should be preventing it from being triggered. If it's already too late for that, then the goals should be protecting users while they are experiencing it and moving users to a more rational state of mind as quickly as possible.

In the Robinhood example, there were many experience and service design decisions that likely increased the severity of the fight-or-flight response instead of lessening it.

The first misstep was the lack of preventative measures—many have argued that the young man should never have been approved to engage with high-risk options he wasn't properly trained to trade. Kearns believed his account was set to a maximum loss of $10,000 and could not understand how he had gotten into so much debt. In the week following Kearns's death, Robinhood vowed to add "additional criteria and education for customers seeking level 3 options authorization."

The second failure point was the interface itself. As Brewster pointed out: "They have slick interfaces. Confetti popping everywhere. They try to gamify trading and couch it as investment." The advertising and UIs of Robinhood encouraged risk-taking and deemphasized the possible consequences. Additionally, an oversimplified display meant there were no explanations, tool tips, or context to indicate to Kearns that the huge negative number he was shown wasn't accurate. Even if the loss had been real, and it is all too possible for Robinhood users to rack up devastating levels of financial losses, there was no next step presented. Nothing to offer a path toward resolving or managing the loss. No direction offered.

This was directly related to the third and most grievous issue, the lack of access to human support to help Kearns understand and resolve his crisis. Due to a spike in sign-ups during the pandemic, the Robinhood customer service centers were severely understaffed. This meant that, according to the Kearns family lawsuit, even though he reached out via email multiple times with increasing levels of "desperation," he was unable to reach a live human, let alone a qualified investment expert to calm him down and explain that his situation was not as dire as it seemed. Instead, Kearns received only automated responses, one demanding he pay $178,612.73 within five days, a misleading communication, which only escalated Kearns's feelings of panic.

Without protection, direction, help, or hope, Kearns's fight-or-flight reaction turned inward, becoming self-harm, and his family suffered a far greater loss than any monetary debt. Brewster said in an interview with the *New York Times*, "There is always personal responsibility, of course. But as a society, I think we owe youth some sort of oversight, and it feels like someone was asleep at the wheel there."

While the causes of suicide are complex and shouldn't be pinned on any single factor, designers and product creators owe their users more care and protection against catastrophic failure, and more support during the aftermath of a major mistake than Kearns was afforded by Robinhood. This is especially true of vulnerable users: young people like Kearns, but also people with low financial literacy, those who are inexperienced with technology, and those with age-related memory issues or mental disabilities. With a focus on prevention, clear guidance, and access to human support, designers can make a big impact in these moments of personal crisis.

Fleeing Toward Help

Anticipating human behavior is an important part of designing human-centered experiences. When it comes to designing for a fight-or-flight response, it's helpful to know that the instinct to run away nearly always precedes the instinct to stand and fight. This human tendency to be gone at the first sign of trouble can be seen in the digital world in how quickly users will close a website or app when things get frustrating or weird. "X-ing out" is the flight response for the digital age.

Once that flight instinct is triggered, you do not want to try to prevent a user from leaving. It's important to be respectful of your users and allow people to engage with your product on their own terms. *No* means *no,* and if you try to force a user to stay when they don't want to be there, like a cornered animal, their flight will quickly turn to fight.

If something in your application has triggered someone to leave, all you can do is wait for them to cool off and reengage with them at another time. To have any hope of successfully saving a relationship with a skittish user, it helps to understand why humans run and the patterns of where they often turn when they are feeling stressed.

THE CASE OF...

THE SCHOOL STAMPEDE

Beth Lieberman had just dropped off her children, ages 5, 8, and 11, at school when she saw the plane crash into the World Trade Center through her car window. She returned to the school as fast as she could. "I just had to get them and then think about the next second, because I just didn't know what was going to happen next." Thousands of New York parents flooded the schools closest to the scene of the attack within minutes of the first crash. Even if it meant heading closer to Ground Zero, the instincts of these parents were to get to their children before they escaped to safety themselves.

Before 2001, most models that city planners used to predict how humans would behave in a mass panic situation, like a terrorist attack, anticipated that people would run directly away from a situation like the one at Ground Zero in New York City. Studies performed in the wake of the attack showed that New Yorkers ran, but they didn't run away, as much as they ran toward the people they cared about. Parents ran to pick up their children from school. People saw others hurt and bleeding and ran forward to help get them to safety. Coworkers who needed to evacuate the building took the time to gather everyone in their team together before heading down the stairs as a group, even as smoke filled their floor. In an exhaustive review of literature on mass panic in the wake of the attacks of 9/11, Dr. Anthony Mawson found "The typical response to a variety of threats and disasters is not to flee but to seek the proximity of familiar persons and places." Mawson referred to these familiar people as *attachment figures*—they might be family, friends, neighbors, coworkers, fellow soldiers, etc.

Overall, human flight-related instincts follow a predictable sequence of behaviors in a crisis:

- Help others in the vicinity escape imminent danger.
- Group up with attachment figures in the immediate vicinity.
- If no attachment figures are nearby, go to them, if feasible, even if it involves risk.
- Flee to a safe place as a group, usually targeting a familiar location or location of additional attachment figures.

This model of behavior during a crisis has major implications for civic leaders, but there are also lessons here for designers of all kinds of systems that handle humans under stress. As you design for a moment of high stress, consider who your users' attachment figures are. Who might they run to for help? Who do they need to ensure is safe? Even if it's not a life-or-death situation, these instincts are still there and can be triggered to a lesser degree during more mundane activities. Maybe a young adult wants to check with his parents before buying his first new car. Maybe a mother wants to look in on her toddler after reading a disturbing news article about childhood diseases.

From a customer service standpoint, human contact can be a powerful tool in managing customer stress. When people reach a certain stress level, they will simply refuse to read the website or chat with a bot. Most often, they will call and insist on talking to a "real human," and the best thing a company can do to calm them in that moment is to answer. To make the best use of the human workforce within any service-oriented system you are designing, you'll want to anticipate these moments and prioritize human intervention. To find high-stress moments along a typical customer journey, look for steps or stages that meet two criteria:

- **High-stakes:** Typically, something where a mistake will be costly or dangerous to the user

- **Low-confidence:** An uncommon or complex activity where a user has little experience performing a task and is very uncertain if they are doing it right

Users perform high-stakes activities, like paying large bills online, without customer service assistance all the time. Because it's such a common activity, they will initiate payments of hundreds or even thousands of dollars without even considering asking for help. Users also do uncommon but low-stakes activities, like enrolling in an email newsletter, without the need for additional help. But if an activity is both high-stakes and unfamiliar, like filing a large insurance claim, that's when they are most likely to seek human support to ensure that they get it right.

Customers also need human support in moments where something critical goes wrong. Giving users an easy path to human connection is essential to reduce panic in those moments and keep people rational. When a bank account is discovered to be overdrawn through fraud, when a medical test comes back positive for a deadly virus, when a news story reports mass violence in a city where a loved one lives, these are the moments when users have an undeniable instinct to reach for human connection—either for expert help or the comfort of loved ones. Identify those possible worst-case scenarios within your own customers' journeys and consider how to put help and human support within easy reach at those critical moments.

De-escalation Techniques for UIs

Often, when you think of de-escalation techniques, you think about something like a hostage negotiator—one person trying to talk the other down from some kind of violent act. However, when it comes to designing for users who are having a fight response, there are de-escalation techniques that designers can incorporate directly into user interfaces to improve your chances of calming angry users.

Much of the literature around methods of de-escalation comes from police-oriented publications, such as the *Journal of Police Crisis Negotiations*, and assumes these techniques will be used as part of a person-to-person interaction. Because of this, some of the proven de-escalation techniques aren't particularly relevant when it comes to designing an online form or digital experience (such as "Use non-threatening body language"). However, there are still many proven de-escalation techniques that can be borrowed by digital designers to improve human-to-computer interactions that involve a human in the fight phase of a fight-or-flight response.

A Warm, Respectful Tone

Nearly every de-escalation training starts by recommending that you speak calmly and respectfully. For this reason, you should avoid using red or all caps text because this can be seen as the digital equivalent of raising your voice. If you're writing content for a system that may be dealing with angry customers, it's important that the language is polite and follows social norms, like apologizing sincerely if a mistake was made. Be careful not to take it too far, however. An overly formal tone can backfire because it implies emotional distance, and the last thing you want is to make the customer feel like you are on opposite sides. Warm and respectful is your best bet for tone of voice.

Acknowledge the Issue and Address It Immediately (if Possible)

The fastest way to deal with a fight response is often to directly address the perceived injustice that triggered it. In some situations, an interface can offer immediate amends without the need for human intervention, like giving the customer a refund, an upgrade, a discount, or some other olive branch to rebalance the equation. It's important when designing these interactions that there are no promises made that can't be kept.

NOTE ANGER AS AN EVOLUTIONARY ADVANTAGE

When you are designing systems that seek to manage human emotions, it's helpful to remember that emotions serve an evolutionary purpose in humans. The emotion of fear alerts people when they are in danger. The emotion of sadness alerts people when they have lost something important. The emotion of anger evolved to alert people to injustice. If you've triggered an anger response in a user, then it's because they perceive an injustice in your exchange. If you can address the injustice that triggered your user's anger and rebalance the equation in a way that seems fair to the user, that's often the fastest way to calm an angry person and avoid that anger from turning to rage. Rage is a mindless type of fight response that takes over when an injustice is seen as a mortal threat so severe that it must be dealt with immediately and at all costs, suppressing even a sense of self-preservation. There is no way to reason with a person who is experiencing rage.

Minimize Distractions

Another piece of advice given to police officers attempting to de-escalate a situation with an angry individual is to try to get the person to a quiet space with minimal distractions. This step allows them to have a focused conversation with that person. From a UI design perspective, this advice might translate to screens decluttered of ads and other distracting messages. For example, rather than a chat box in the bottom corner of a larger page, take the user to a page dedicated to the chat conversation.

Use Plain Language

Clear communication is an important part of de-escalation. Avoid jargon and abbreviations that may not be meaningful to the user. As is recommended to professional negotiators, use short words and simple sentences.

Ask for Details

In many tense situations, it's important for people to feel they have had a chance to tell their side of the story. Having a UI gather relevant details can reassure users that their situation is being taken seriously. If details are gathered, any follow-up communications must demonstrate awareness of the information that was previously shared by the user. One important note about follow-up questions that comes straight from the negotiator's manual: when possible, avoid asking "why" questions to someone who is upset, because it often triggers defensiveness and escalates tension.

Be on Their Team

The language used should express a sincere desire to help the user and show empathy for their situation. The more stressed that users become, the more they fall into "us versus them" mentalities. Strategic use of plural first-person pronouns like "we" can imply that you and the customer are on the same team, which can disarm a fight response.

A Cooling Down Period

The final way to deal with a fight response is to disengage temporarily, which is essentially the equivalent of adding a flight response to your system. This method gives the person a chance to cool down and come back to the discussion with a less aggressive frame of mind. It must be done in a way that is respectful of the user. For example, if an AI system senses a user is becoming increasingly upset while waiting for a chat representative, it might offer a way for the chat rep to reach out to the user at a later time that is convenient for the user.

Coral is a commenting and community management tool built by Vox Media that attempts to elevate online engagement. It leans heavily on UX research to find ways to reduce fighting, abuse, and bullying, while increasing productive online dialogue in news media's comment sections.

Slowing users down is a surprisingly common theme in their toolkit for better online debates. The first way they slow folks down is just to require them to create an account to comment on an article. During the sign-up process, members must agree to the community code of conduct. After sign-up, the community rules are continuously and prominently displayed at the top of the comments for every article, as shown in Figure 4.2. Studies have shown that just showing someone community rules before commenting is a surprisingly effective way to decrease inappropriate posts measurably. The moment of pause and the reminder of the rules seem to prime commenters for better behavior.

But, of course, reminding people of the rules doesn't always guarantee discussions will remain cordial when topics get contentious. Advanced moderator tools allow administrators to put particular commenters in a temporary "time-out" or pause commenting on the entire post when community members get too riled up and discussions turn to fights. These cooling down periods have been shown to significantly curb fighting among community members.

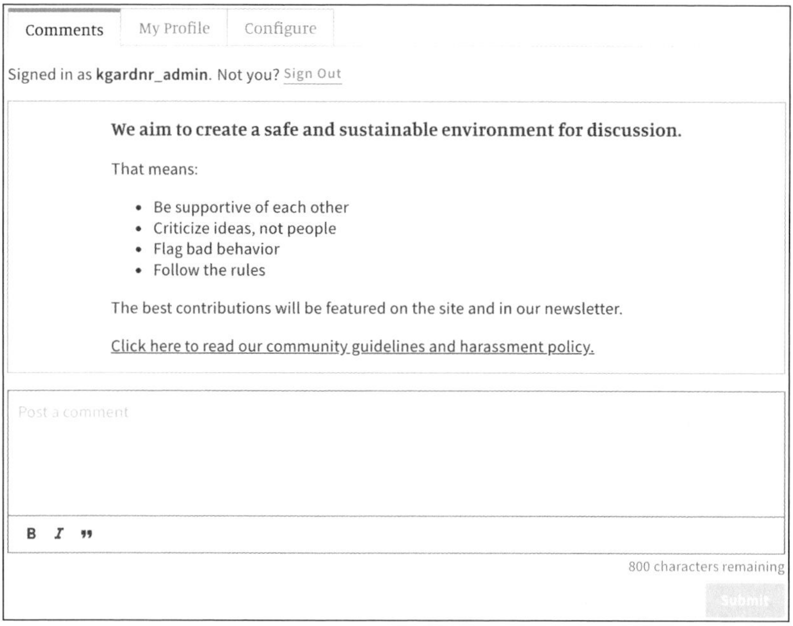

FIGURE 4.2

The Coral Project by Vox Media features highlights from the community guidelines at the top of each commenting section to remind users of what good behavior looks like.

In a similar vein, Twitter has added a couple of automated prompts to nudge users toward better behavior, as seen in Figure 4.3.

The first prompt, released in September 2020, detects if someone is about to retweet a story without first opening the article. Rolled out shortly before the contentious 2020 U.S. election, the intent of the feature was to slow down the spread of misinformation. According to Twitter, the feature increased the number of people who opened the article before retweeting by 33% and caused some to decide not to share after all.

A similar tactic was used in a second feature rolled out several months later, which used natural language processing to detect "potentially harmful or offensive language" and prompt users to review their tweet before posting. According to TechCrunch, Twitter reported "34% of people revised their initial reply after seeing the prompt, or chose not to send the reply at all."

FIGURE 4.3
Automated Twitter features intend to slow the spread of misinformation and reduce offensive language.

When designing any system that a user may use while angry, look for ways like these that give users time and space to rein in their tempers and re-center themselves around what really matters. And, be sure to balance any such feature with safeguards to ensure that important topics, especially topics that you fear may be censored by those in power, are not inadvertently silenced.

Getting Unstuck

If the brain judges that it is impossible to either escape or successfully fight off an attack, *freezing* is the body's last-ditch effort at survival. At the extreme end of this reaction is *tonic immobility*, which has long been observed in animals such as the opossum, famous for "playing dead," as well as in humans caught in extremely traumatic situations.

In the less extreme context of a consumer experience, behavioral marketers have noted a phenomenon they call *choice paralysis*. There is a well-known study from 2000 that illustrates this through the sale of jams at a grocery store. When presented with 24 flavor options, shoppers bought significantly less jam than when there were only six flavors on offer. The shoppers found too many choices to be overwhelming. They couldn't make a decision, so they purchased no jam at all.

A Single, Clear Directive

When stress levels rise, having an abundance of choices and no clear direction is a recipe for a freeze response. On TV when someone has a heart attack, there's usually a character who yells, "Somebody! Call 911!" However, CPR instructors will tell you, this is not the ideal way to handle it. Even if there are ten people standing around, it's not unusual for all of them to be frozen with uncertainty, and no one actually moves to make the call. (This is a well-studied phenomena called the *bystander effect*.) It's much better in that moment to point to a specific individual, make eye contact, and say, "You. Call 911." This process almost always ensures that it will get done.

That kind of clear, specific, singular direction is what users need in a moment of crisis. Let's say they've just gotten some very bad news through your app, like their identity has been stolen and is being used by hackers for nefarious purposes. They are obviously distressed about this, but they aren't sure what to do about it. The longer they stay in that moment of helpless indecision, the more their cortisol level rises, making rational thought harder and harder.

This is not the moment to give someone a bunch of choices, especially on the very first step. Instead, give them a single, clear path forward. If you do give them choices, limit the number of options as much as possible. It's far better to create a flow that offers a series of short decisions than a single page with many options. Even if it adds more clicks, it will help the user get "unstuck" and build positive momentum.

Least-Worst Decision-Making

Choice paralysis most often occurs when the outcomes of various choices are essentially the same. When one outcome is obviously good and the other is obviously bad, the decision comes to mind intuitively. But when options are too close, the subconscious kicks the decision to the conscious mind to make a determination using a more analytical approach. However, if the brain is flooded with cortisol from a stress response, that logical thinking may be compromised, making the choice even harder to make.

What's ironic about choices like this is that it's actually the lack of difference of impact that makes the choice so hard. Your brain has to put a lot of effort into making a choice where the negatives and positives of each side balance to essentially the same total. In cases such as these, Klein suggests simply flipping a coin to make a decision so as not to waste the mental energy.

But not all tough decisions are about topics as frivolous as picking a flavor of jam. In battle, soldiers must sometimes make impossible decisions between two bad outcomes. In circumstances such as these, a flip of a coin is an unacceptable choice model. In their book *Conflict*, Neil Shortland, Laurence Alison, and Joseph Moran, researchers of military decision-making, explore the process of *least-worst decision-making*. For an officer to freeze up in these circumstances is often the truly worst option since time ticking away quickly reduces the choices available to the troops to minimize casualties.

Shortland et al. found that goal-based thinking is one of the best ways to overcome decision inertia. For soldiers, a strong sense of core values was essential in order to avoid choice paralysis. Moral mandates like "never leave a fellow soldier behind" serve to re-clarify goals in otherwise murky situations such as when plans fall apart in the heat of battle.

Occasionally, however, soldiers face choices where two core values conflict, such as loss of life of innocent civilians versus loss of life of fellow soldiers. No matter what they choose, a core value will be violated. In these cases, one of the best strategies for avoiding a freeze-up will be to simply stop trying to make a "good" decision and accept that their task is actually to choose the least-worst option in order to minimize harm. In the moment, this can seem counterintuitive, but it is, in fact, a highly effective way to break a mental freeze and minimize losses. As the authors put it, "The quest for a positive outcome creates a negative outcome, whereas the acceptance of a negative outcome creates a positive outcome."

As a designer, you've likely seen the importance of problem framing in your own work. If you are creating interfaces or systems designed to help users make choices, consider how you are framing those choices within your interfaces.

First, remember that, as Klein said, 90% of choices will be so obvious that your users won't even consider them choices. Don't belabor those in your interfaces. Aim for an interface that is as effortless as the "decision" itself. An example of this might be setting default choices for fields that are nearly always answered the same way (while still allowing them to be easily edited) instead of having users put in the information themselves.

For the other 10% of choices, help your users avoid choice paralysis by reminding them of their goals and core values. (More on defining core values of diverse user groups in Chapter 5, "Reasoned Reaction.") If your users regularly face the unenviable task of having to make a least-worst decision, avoid framing questions with phrases like, "What do you want to do next?" since no one *wants* to make a least-worst decision. This type of framing will only prime them to freeze up and be unable to make a decision. Rather, frame the question in a way that helps them understand their job is to choose the option that goes with their values (or the values of their organization) and causes the least amount of harm.

Design Techniques for Fight or Flight

The fight-or-flight response is an ancient survival mechanism that is sadly ill-equipped to help with many issues in the modern world. So many of the devices developed to make people's lives safe and easy rely on the user to think and behave like a modern human with fine motor skills and a fully functioning logic and language center. But when users panic, that veneer of civility gets stripped away momentarily. If you, as a designer, don't fully embrace how truly different people become in that moment, you may find the thing you designed to help someone will instead become the very thing that blocks them from getting what they need most.

THE MISSING NINE

It was a typical summer morning, and Daniela was tidying up the upstairs bedrooms as her youngest son followed her from room to room. She left her bedroom for just a moment to grab something from the hall closet and when she returned a minute later, the room was empty. Her son had disappeared.

Then she noticed the open window, the screen no longer in the frame.

"I knew immediately what had happened," she said. She ran over to the window and looked down to see her four-year-old son, two stories below, lying still on the concrete patio.

She ran down the stairs screaming her son's name and grabbed the cordless phone on the way through the kitchen. She rushed to her son's side. He was still breathing but unconscious, not responding to her at all.

She stared down at the phone in her hands. She knew she needed to call for an ambulance.

"My fingers were frozen. I couldn't press the buttons. I looked at the numbers, and they were just a bunch of symbols. *I couldn't find the nine.*"

By this time, her older son had heard her screams and came to see what was wrong. She turned to him and said, "You have to call 911." She held out the phone to him with a trembling hand.

It was her eight-year-old son who placed the call. He saved his brother's life that day.

This is a true story that a colleague shared with me when I was first starting to research this book. Thankfully, the story had a happy ending. After a few days in the hospital, Daniela's son was able to go home, and he's since grown up happy and healthy. There were no lasting side effects from his fall.

Daniella's son fell over a decade ago, and in all that time, phones have become even more difficult to dial. It typically takes about five to eight clicks on a smartphone to close out of your last app and navigate to the dial pad. In Figure 4.4, you can see a typical sequence.

FIGURE 4.4
The four screens I navigate through to get to my iPhone dial pad.

More than 80% of 911 calls are made from mobile devices. There are emergency shortcuts built into most smartphones, but they tend to be hidden and unintuitive. As far as these things go, the iPhone "SOS" feature is probably the best of the major market solutions, but none are as intuitive or visible as they should be for a feature on which someone's life may depend.

DESIGN QUICK LOOK
SENDING OUT AN SOS

The iPhone "Emergency SOS" feature, shown in Figure 4.5, gives users several shortcuts for emergency services calling. Users can pinch the volume and side buttons to trigger a screen where they can swipe to place the emergency call, or they can rapidly click the large side button five times to trigger the call. The rapid clicking solution seems to be the most human-centered design because it uses the largest button on the phone (reducing the need for fine motor skills), and that kind of rapid "panic clicking" is an achievable motion even under an adrenaline rush.

FIGURE 4.5
iPhone's Emergency SOS features are more fully fleshed out than most, but many users are unaware of how to activate them.

As soon as the emergency call is placed, automated text messages with your current location will be sent out to those marked as your emergency contacts. They will continue to receive location updates via text until you disengage SOS mode. Additionally, for users with an Apple Watch, if the device detects a hard fall, it will automatically trigger the SOS feature and call for an ambulance.

Unfortunately, there are two major drawbacks to this feature's design. The first is that it's too easy to trigger them in a purse or pocket. Emergency services have noted a definite increase in 911 "butt dials" since the feature was implemented. (In one California county with a major Apple repair center, as many as 20 false calls a day were received according to a 2017 report.)

The second drawback is that many people don't realize the feature exists or remember it's there when they need it during an actual emergency. The phone needs stronger visual indicators of which button to press, the shortcut should be standardized across brands, and there should be an educational campaign similar to when 911 was first rolled out to inform the public of the new standardized shortcut for all mobile devices.

Crisis Controls for Visual Interfaces

A panic response is fairly rare, but there are some circumstances where it can be anticipated that a user is likely to be in or near a fight-or-flight state. For instance, Allstate added a crash detection feature to their app that uses accelerometer data and app notifications to let users know they aren't alone in the moments directly following a crash, and that Allstate can help. The first thing offered by the app is a shortcut to dial 911. For users who don't need emergency services, they are given shortcuts to the two most common tasks users turn to Allstate for in the moments after an accident: ordering roadside assistance or filing a claim. As part of designing this set of features, Allstate developed a special set of crisis controls that are specially designed to accommodate someone in fight-or-flight mode.

To create crisis controls for your own product, consider the following:

- Enlarged controls
- Full width
- Increased legibility
- Linear decision path with a reduced number of choices

Enlarged Controls

Buttons should be sized up significantly from their normal size to accommodate clumsy fingers and shaky hands. Also, leave white space around those enlarged buttons to accommodate the impulse variability that was discussed in Chapter 2, "The Startle Reflex."

The limiting factor here is that you don't want to make buttons so big that users can't immediately see the action button when the page first loads. You want to provide a clear, instantly visible path forward on every screen, ideally with no scrolling needed.

Full Width

When designing for a mobile device, style your buttons to cross the full width of the screen. This design increases the tap area and enables users to reach buttons with the thumb of either hand. In an emergency situation, it is more likely that a user will be using an interface one-handed, either due to an injury or the need to use their other hand to stay safe.

Increased Legibility

You should also ensure that your text is extremely legible through word choice, color contrast, and fonts. For the words themselves, use short, simple, common words that can be read intuitively by the brain. (See the "Dare You Not to Read This" sidebar later in this chapter for more on intuitive reading.) Additionally, consider the color contrast between the text and the background. Free, online contrast checkers can be used to check the amount of contrast between colors, which is usually expressed as a ratio with 1:1 being the same color and 21:1 being highest possible contrast—black to white. Web Content Accessibility Guidelines (WCAG) recommend using at least a 7:1 color contrast ratio to be accessible to a wide range of users.

In terms of font choices, there isn't one specific font that is the best for readability, as much as a list of characteristics that are important within any font that is appropriate for your brand and audience. The Readability Group, an organization dedicated to optimizing typography legibility through "data not anecdote," has found through their usability research and work with neurologists that it's important to choose fonts that emphasize the following information:

- Minimize the occurrence of imposter letters, characters that can be easily confused with each other such as 0 and O, or capital I, lowercase l, and the number 1.

- Minimize mirroring in letter shapes, ensuring that characters like lowercase b, d or p, q have subtle differences that prevent them from being exact mirrors when flipped. This is especially helpful for readers with conditions like dyslexia, but helps all readers.

- Allow letters to be easily distinguishable from each other through letter shape, line spacing, and differing heights between capitals and *ascenders*, the parts of the letter that stick up on letters like lowercase b's and h's.
- Adjust the spacing between letters (sometimes called *kerning*) to ensure that letter pairs aren't confused for single letters such as "rn" appearing as an "m."

DARE YOU *NOT* TO READ THIS

Some kinds of reading are completely intuitive and even involuntary for most users. Let's say that you are looking at a laptop and this flashes on the screen:

Cat

If you are fluent in reading English, you actually can't prevent yourself from reading a word that is short, common, and simple. In fact, I challenge you to look at the words in the following sentence *without* reading them.

Lucy is going to the park and she is taking the dog for a walk.

If you are literate in English, you likely find it impossible to prevent yourself from reading those words. It happens completely intuitively and is outside of your conscious control. However, the longer and less common the word, the less likely that someone will be able to read it intuitively. Since people rely more and more heavily on their intuitive skills as their stress levels rise, stressed users will have a significantly better chance of being able to use your product if they can lean on their intuitive reading skills when using your interface. To enable this, keep words in your UI short, simple, and succinct.

Linear Path, Reduced Number of Choices

Whenever possible, make the first action just a simple "start here"–type button to help users overcome any freeze response they may be experiencing. From there, create a linear path for users with simple choice architecture, presenting only two or three options, or binary yes/no choices at each step along the way.

Controlling a Crisis Through Voice Controls

George Salazar is a NASA engineer who was integral in the development of the Space Shuttle flight experiment investigating the operational effectiveness of voice control of a spacecraft system. "We look at speech control as a third pair of hands," George explained. "Speech control is more for when your hands and eyes are busy." The most common use case of voice controls in space would be an astronaut telling an external camera to "pan right" or "zoom in" while the astronaut executes remote repairs using a robotic arm. Using voice commands allows them to adjust their view of the task while keeping their hands on the robotics controls and their eyes on their repair work in progress.

But voice commands also serve an important role in contingency design within the shuttle. The first scenario is what George calls an "incapacitated crew." If the oxygen levels get so low that the crew is immobilized or if they're injured in a way that restricts their ability to use physical controls, the voice commands are available to them as a backup. The second emergency scenario is called the "dark cockpit." Just as the name implies, if power is lost to the main visual display bank and controls in the main cockpit, the system can still read out statuses to the crew and the crew can give the system verbal commands in response.

George pointed out that voice command, while an important redundancy, isn't ideal for high-stress exchanges, due to current technical limitations that don't play well with the human stress response. "When things go wrong, sometimes people start shouting," but getting louder doesn't necessarily help computer-based systems hear commands better. "From an electronic standpoint, you saturate the microphone signal input to the speech recognizer. Speech gets distorted, and makes it harder for the system to understand."

Another drawback is that when stress levels rise, people tend to speak faster and reduce the length of pauses between words and sentences. A language processing system *listens for* and *relies on* those pauses to help determine what words are spoken. When a user gets stressed, they may complete their instruction to the computer and then immediately begin talking to another human in the cockpit without an appropriate pause or putting the system into a standby mode. Additionally, things tend to get noisy during emergencies, and the system can't always separate spoken commands from background noise. Lack of appropriate pauses, forgotten standby commands, or excessive background noise can cause issues with *end of speech detection*. Without a clear end detected, the system will remain in listening mode, gathering irrelevant or incorrect commands, slowing down system reaction time, and generally making a muck of things.

With currently available technology, there aren't great work-arounds yet to these limitations, which is why voice commands remain a tertiary control option for most critical tasks in high-stress situations. Jonathan Bloom, a conversation designer at Google, confirmed that these limitations persist in consumer technology, not just on spacecrafts. "Voice interfaces can absolutely be a challenge when callers are upset," he said. "Voice is great for short-cutting through a lot of menus, but not if the caller is agitated or in a chaotic environment." If you are considering voice control options for stressed users in your app, it can be a life-saving contingency for tasks like calling 911 if a user is physically trapped or critically injured after an accident, but it should probably not be considered a primary path for most high-stakes tasks, more of a critical backup.

Critical Information: Fight or Flight

Human behavior under the effect of the fight-or-flight response varies widely, but it does follow some patterns that allow designers to create systems that can reliably assist users in a crisis.

How does the fight-or-flight response work and what is a designer's role in managing the effects?

The fight-or-flight response is composed of three sequential steps:

1. Flight
2. Fight
3. Freeze

A designer's role is to minimize the negative effects of the response through preventing the fight-or-flight response from being triggered, guiding the user toward a productive path out of their panic, and providing access to human support in moments of true crisis.

How should designers handle users who are trying to flee from a product or experience they have designed?

People don't just run away from danger; rather, they run toward people they care about. As a designer, one of the best things you can do for someone in a flight response is help them get to a human who can help them through their moment of panic. To do this, be deliberate in placing lifelines to help that are easily accessible to users right in the moments they need that help.

What de-escalation techniques can calm a user in fight mode?

The fight response is triggered when someone is feeling trapped, either physically or through circumstance. Anger is an emotion that comes in response to a perceived injustice. To de-escalate after a fight response has been triggered, UIs should do the following:

- Use a warm, respectful tone.
- Acknowledge the issue and address it immediately, if possible.
- Minimize distractions.
- Use plain language.
- Ask for details.
- Show that you are on their team.

And, if all else fails, pause the interaction and allow the user to cool down.

How can designers help a frozen user get unstuck and take positive action to address their stress?

To help a user get unstuck, give them a single, specific action to take. If someone is faced with having to make a decision with two bad choices, instead of asking them to make the best choice, frame the choice around picking the option that causes the least amount of harm.

What are specific UI design techniques that make interfaces easier to use when someone is in a fight-or-flight mindset?

The following design techniques can make interfaces easier to use when someone is in a fight-or-flight mindset:

- Enlarged controls
- Full width in mobile
- Increased legibility
- A linear decision path with a reduced number of choices

Go to the Source

The Wisdom of the Body: A book by Walter Bradford Cannon in which fight or flight was first coined, 1932.

"Understanding Mass Panic and Other Collective Responses to Threat and Disaster": A review of available studies by Anthony R. Mawson, 2005.

"A Practical Overview of De-Escalation Skills in Law Enforcement": An article by Janet R. Oliva, Rhiannon Morgan, and Michael T. Compton, 2010.

TED Radio Hour: "Reframing Anger": A podcast episode hosted by Guy Raz, 2019.

Hidden Brain: "The Logic of Rage": A podcast episode hosted by Shankar Vedantam, 2020.

"When Choice Is Demotivating: Can One Desire Too Much of a Good Thing?": A paper about jam and choice paralysis by Sheena S. Iyengar and Mark R. Lepper, 2001.

CoralProject.net Research Collection: Vox Media has gathered a large collection of UX research on creating a positive space for online discussions and communities which can be found at https://coralproject.net/research/.

Conflict: How Soldiers Make Impossible Decisions: A book about least-worst decision-making by Neil Shortland, Laurence Alison, and Joseph Moran, 2019.

Reasoned
Reaction

Nearly two decades before he took the first step on the moon, Neil Armstrong was a fighter pilot in the Korean War. While flying a skirmish in his F-9F Panther fighter bomber, part of his wing was sheared off by an anti-aircraft cable. Through some truly spectacular flying, Armstrong was able to keep his plane in the air. By maintaining a high rate of speed, he kept just enough lift under his damaged wing to stay in the sky and escaped the battle. But once out of the line of fire, he quickly realized he wouldn't be able to land. If he slowed down at all, his lopsided plane would go into an uncontrollable horizontal spin. His commanding officer instructed him to fly back toward base, eject near the camp, and let his plane crash into the ocean.

Armstrong had only basic training in the use of the ejection seat on his model of plane so while flying a severely damaged aircraft over enemy territory, he pulled out the instruction manual and reviewed the ejection procedures shown in Figure 5.1.

Luckily, the instructions were well designed—broken into simple steps and even illustrated. Armstrong was a quick study, had nerves of steel, and was able to execute the ejection flawlessly. He parachuted safely into a rice paddy just outside the American base.

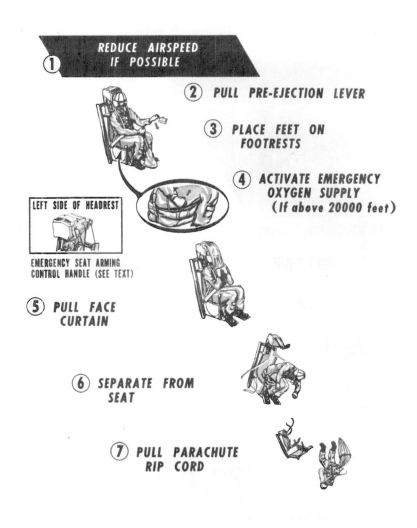

FIGURE 5.1
An excerpt from "Flight Handbook, Navy Model, F9F Aircraft" dated May 1, 1954, showing Armstrong how to eject from his pilot seat.

Steps, systems, structure, and process are the backbone of designing for rationality under stress. When cortisol and adrenaline are running high, a user's capacity for higher-order thinking is significantly reduced. Therefore, the tools that are most helpful are like those instructions, breaking complex tasks into simple, sequential steps.

This process takes the burden off the user, reducing the complexity and number of decisions to be made so they can focus on the critical tasks that have to be addressed in the moment. When a user has clarity of purpose and a process to follow, it allows them to put their stress-induced hyperfocus and drive to good use, keeping them on the tasks that were predetermined to be the right steps.

As we discussed in Chapter 3, "Intuitive Assessment," intuition doesn't follow any kind of well-ordered process. It jumps to conclusions and leaps ahead. There are times when designers want to support intuitive decision-making in their users and times when they want to steer users toward a more rational approach. It is environments with high variability, high complexity, or low familiarity where intuition can do more harm than good. In these circumstances, users often need help avoiding missteps, mistakes, and bias traps. That's where rational design strategies can make a positive impact.

In this chapter, we'll explore specific design strategies for helping your users succeed in unpredictable or unfamiliar environments where a rational approach is needed. We'll explore these key questions:

- How can designers help stressed people who are forced to deal with complexity they have no familiarity with?
- How can designers help professionals who work in unpredictable environments make the most of their expertise?
- What types of systems can protect best against harmful bias?
- Since no designed system can truly be without bias, how can designers ensure that the bias in the system reflects the morals of the communities it is meant to serve?

Zero Familiarity

There are many situations where novices or consumers have to jump in and quickly build expertise in a particular topic. Unfortunately, high levels of stress often make learning new things more difficult, which is true for both positive and negative experiences—from buying a car to filing taxes. But perhaps one of the worst combinations of experiencing high stress and needing to learn a complex new space is when you or someone you love is diagnosed with a potentially deadly disease. Designers who design patient services are constantly working to perfect the art of communicating complex information to people who are simultaneously trying to process life-altering events.

At Michigan Medicine, a research hospital on the campus of University of Michigan, clinical trials are a regular part of the treatment process. Most trials performed are Phase 3 clinical trials where slight tweaks to the standard care are tested. These types of tests rarely make headlines, as they typically only make incremental improvements of a couple of percentage points, but those gains over time add up to a significant benefit for patients. For instance, the survival rate for childhood leukemia has risen from 48% to nearly 90% over the last 50 years, largely due to steady improvements made through phase 3 trials.

Undoubtedly, these types of clinical trials are a critical part of advancing medicine. But if it were *your* child with leukemia, would you be willing to enroll them in an experiment? It's a difficult decision to make. And if the parents do agree to sign up, it's critical from an ethical standpoint that those parents understand what it is they are agreeing to.

In an effort to better understand the experience of patients and their families who participate in these trials, the office of pediatric oncology at Michigan Medicine surveyed the parents and guardians of children with cancer who received treatment at their hospital. The results highlighted these trends with parents who agreed to allow their child to participate in a phase 3 trial:

- Only half of the parents could explain the goal of the clinical trial.
- At least 66% of them were unaware they had an alternative to participating in the trial.
- One in five parents didn't even realize their child was enrolled in the trial.

The research team knew something needed to be done to address the poor communication with parents and patients around the clinical trial enrollment. Dr. Laura Sedig approached Michigan Medicine's UX team to see if they could possibly help with a digital solution.

Jackie Wolf, the lead designer and UX researcher for the project, immediately understood the issue as soon as she saw the dense, 30-page consent disclosure that was handed to parents during these conversations. It was full of medical jargon and incomprehensible charts, such as the one shown in Figure 5.2.

FIGURE 5.2
This chart, meant to illustrate how a phase 3 clinical trial works, is from the original consent disclosure agree form that was eventually redesigned.

Although a doctor would often try to talk a patient through the documentation, the success of those conversations was hit or miss. Parents had often been awake, in the hospital for days on end, while their child endured a battery of tests, with the end result being devastating news that their child had cancer. Terrified and sleep-deprived, they were in a poor state of mind to process new vocabulary like "randomization" and "branches of the experiment." They often didn't feel that they could question the doctor, and the accompanying materials did more harm than good when it came to comprehension.

Jackie wanted to get this design right. She had deep empathy for the parents in this terrible situation, and she also knew the medical importance of the clinical trials. So she dove deep into the existing research available on the subject. After reviewing dozens of studies with titles like "Stress and Selective Attention: The Interplay of Mood, Cortisol Levels, and Emotional Information Processing," Jackie created a list of goals and guidelines for her work:

- Combine written and visual content.

- Break content up into small, bite-sized pieces.

- Create an interactive experience that allows users to go back and revisit information if needed.

- Use simple, conversational, empathetic language.
- Model behavior that the user should emulate in real life.
- Make the information relatable, to help users see themselves in the content.

You can apply these same principles to your own work when designing any product that will be used by regular people who are in high-stress, highly complex situations.

Combine Written and Visual Content

There is one medium that does an especially good job combining written and visual communication: comics.

Comics may not be an expected format for conveying medical information, but it's proven to be a surprisingly effective one. Jackie's comics don't feature superheroes or cats who hate Mondays. Instead, they take the complex concepts of clinical trials and break them into single-panel storyboards showing a conversation between a doctor and two parents, as can be seen in Figure 5.3.

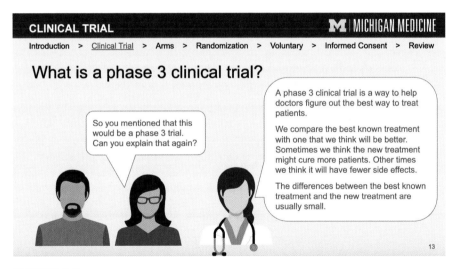

FIGURE 5.3
Michigan Medicine's Clinical Trial Consent Disclosure Agreement Explanation used a comic book format that leveraged all of the best science on cortisol and learning.

Bite-Sized Pieces

The comic format also does a great job of leaning into the second guiding principle: break content up into bite-sized pieces of information. The format of speech bubbles serves as a natural limiter, encouraging visual "chunking" of the information.

Make Information Interactive with the Ability to Revisit

When it comes to interaction and learning, studies showed that even just turning pages helped users stay more engaged with the information they were trying to learn when compared to just sitting back and watching a long movie or presentation on the same content.

Parents were presented with the comic on an iPad. Clicking through each page of the comic allowed them to absorb the information at their own pace. Additionally, the navigation along the top allowed parents to click back into sections they had previously read through to review them if they needed.

Use Simple, Conversational, Empathetic Language

Since this was such a content-heavy experience, in order to get the words just right, the team brought in Nancy Nilles, a medical writer who specialized in breaking down complex medical issues into plain language. Having lost her own mother to cancer several years earlier, Nancy was familiar with the experience of having a loved one in the medical system: "It's such a large entity. It has its own energy, and you can just be kind of swept up into it. It can be really disorienting." Nancy was able to draw heavily on her personal and professional experience to dial in the right tone for the experience. "I tried to use a lot of sympathetic language and a gentle tone."

Here are some examples of how Nancy achieved that gentle tone, as shown in Figure 5.4:

- Acknowledge the difficult situation.
- Avoid a clinical tone or jargon, keeping it conversational.
- Don't assume that the user either knows something or doesn't know something.
- Signal that it is OK not to understand something and to ask questions.

- Share common questions and concerns that other parents have expressed in this situation.
- Speak directly to users with second-person language (for example, "you might" instead of "parents might").

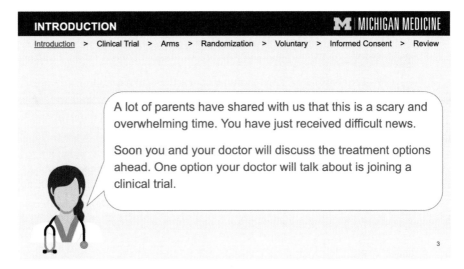

FIGURE 5.4

In this panel that comes early in the experience, the doctor character explicitly acknowledges how difficult the situation is for the parents before jumping into the bulk of the content.

Nancy ran all of her work through online evaluation tools to ensure that the content never exceeded an eighth-grade reading level. These efforts were especially important to protect the populations who were most vulnerable to low health outcomes—those with low levels of completed education or those who spoke English as a second language.

Even with a focus on accessible language, Nancy knew that getting parents comfortable with some new vocabulary was important to empower them in future discussions with hospital staff. She was careful to define important terms early in the experience. "When I first received the [content] deck, it had some pretty technical terms high up." Uncommon phrases like "arm of the experiment" were only defined in a glossary in the back. Nancy quickly realized, "Those first slides didn't mean anything to you if you didn't already know all the terms," so she ensured that medical words were always defined when they were first used.

Additional efforts to simplify language included avoiding acronyms and including lots of context clues along the way to help users remember what new medical terms meant. Nancy knew that in such a high-stress, high-complexity experience, the user's working memories would be limited and retention of new knowledge would be low.

> **NOTE** ARE ACRONYMS BAD UX?
>
> You generally want to steer clear of acronyms in experiences designed for nonexperts. However, some acronyms are used so often that the "real" term may not be recognized by a layperson. For example, if you heard that someone had high levels of tetra-hydrocannabinol in their system, you would be less likely to know that it meant they had imbibed marijuana than if you heard they tested positive for THC. Other examples include cardiopulmonary resuscitation (CPR) and automated teller machine (ATM). In these cases, the full term is less intuitive than the acronym.

Model the Behavior

The comic-style layouts helped address another issue Jackie had found through her research. Often, parents felt that they shouldn't be questioning their child's doctor, when, in fact, that sort of back-and-forth dialogue often results in significantly better understanding and outcomes. So the comic intentionally showed parents modeling the behavior of asking questions, which gave parents explicit clues that this was an expected behavior for them.

In one particularly impactful panel shown in Figure 5.5, the father asked: "Does that mean you would be experimenting on my child?"

This discomforting question about experimentation had come up often in Jackie's early user feedback sessions. She and Nancy chose to include the question using that same verbiage to address the concern head-on. Modeling this type of "real" question was important to empower parents to ask not just clarifying, but also challenging questions.

Jackie was also deliberate in showing two parents, one of each sex, both of whom asked questions. She felt this was important because a study she had found showed that mothers tended to carry the burden for asking questions, remembering details, and making final treatment decisions. She wanted to model a situation where both parents worked together to understand their child's treatment options.

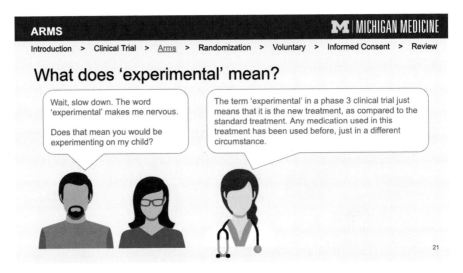

FIGURE 5.5
The comic intentionally models parents asking difficult questions of the doctor. This comic addresses common questions head-on and demonstrates the type of open dialogue the experience hopes to encourage in real parent-doctor interactions.

Make the Information Relatable

After they had sketched out the initial storyboards, Jackie's team looked at a variety of illustration styles to bring the comics to life. Early cartoon-style drawings, seen in Figure 5.6, didn't seem to take the situation seriously enough.

The team finally settled on the icon-like illustrations. These figures were drawn deliberately vague. The goal was to create representational figures that a wide variety of parents would be able to relate to.

All of these research-backed techniques—combining written and visual information; chunking up the content; creating an interactive experience; using simple, conversational, gentle language; modeling behavior; and making it relatable—culminated in an interactive tool that takes better care of these parents during an incredibly difficult moment. In some ways, all of these techniques are fairly basic and "just good design," but in this particular combination, used with such care and intention, they create the best possible learning environment for parents under enormous amounts of stress. Details matter when stakes are high, and these principles are excellent models to follow when you create experiences for your own stressed-out users.

Early Illustration Exploration

Final Illustration Style

FIGURE 5.6

An early prototype illustration in a more "cartoon" style versus the more icon-like final figure designs.

Experts in the Unpredictable

The Michigan Medicine case study focused on parents, most of whom were not medical experts. But what about those designers who are creating products or systems for experts? Many experts are forced to deal with complexity in high-stress situations and in these circumstances, oversimplification can actually cause more problems than it solves. A very different design approach is needed for these types of users.

Samantha Szymczak is a human factors engineer who designs applications that are used by the intelligence community. Sam designs tools to be used by senior data analysts at organizations like the CIA who are tasked with poring through thousands of pieces of data to try to circumvent imminent attacks on the United States or their troops abroad.

These people have an acute awareness of their own bias. In Chapter 3, Kahneman and Klein specifically listed "intelligence analysts" as a field so unpredictable that intuition could not be trusted, which is a known and accepted reality in that field. When analysts make a report, their top goal is to provide a complete and coherent assessment of the situation with as little corruption as possible from common mental traps such as confirmation bias, availability bias, or poor framing effects. These are clear and present dangers when working with messy, real-world data that include partial data sets, bad sources, and *unknown unknowns* (things you don't know that you don't know.)

Sam and her fellow designers have a series of design tenets that guide their work when creating software and tools for expert users in a high-complexity, low-predictability environment:

- Assign tasks appropriately.
- Expose the bias.
- Prioritize ruthlessly.
- Support the decision.

Assign Tasks Appropriately

"Assign the computers the work the computers are good at and the humans the work the humans are good at," Sam advised.

Humans are great at making connections, spotting obscure patterns, and pulling relevant insights. But humans, as a rule, are terrible at accurate, detailed recall. Computers, on the other hand, have perfect accuracy when storing and retrieving information. And while computers are great at spotting patterns, they can't determine which patterns are most meaningful without a prescribed algorithm, nor can they apply insights that a human hasn't already programmed them to use.

An analyst may be working on a report for months. Over time, they may remember the relevant facts, but no longer be exactly sure where they learned each one, how old or reliable the source was, or what all the context was surrounding the memorable data point. In a well-designed system, the computer and human strengths work together—each insight the human pulls out and notes in their report is linked through the system to the original source of that data so that the context is easy to retrieve.

Whenever you design a process where the computer is meant to supplement human memory or ability, consider carefully the strengths and weaknesses of humans and machines. This is a huge hurdle for the transportation industry currently. The development of autonomous capabilities in commercial trucks and consumer vehicles is in an awkward stage where humans and machines are being tasked with the wrong jobs because of limits in the technology. Humans are being asked to watch a machine perform a mundane task for long periods of time and remain fully vigilant throughout, catching any errors the machine makes and jumping in to correct course in the seconds or sometimes split-second between when the system should have acted and when it's too late to act.

As mentioned briefly in Chapter 1, "A Designer's Guide to the Human Stress Response," boredom and monotonous activities cause hypostress (*hypo-* coming from the Greek root meaning "under" or "below"). And just like all stress responses, hypostress triggers unpleasant feelings that drive humans to take action to relieve that stress. In the case of relieving hypostress, this usually means that people find something else to do to occupy their attention or bring a sense of purpose. Most of the deadly accidents caused by automated systems occur because monitoring the system was so boring that the human tasked with the job went and found something else to occupy their attention. Infamously, the first person to die while his Tesla was in autopilot mode was reported to have been watching a Harry Potter movie instead of monitoring the road. The car failed to distinguish a white truck from a bright sky, and the man was too distracted to catch the system's mistake in time to avoid a deadly collision.

Designing a successful partnership between humans and machines is a delicate balance and requires a firm grasp of the limits of each. For Sam and her work designing for intelligence analysts, the relevant human limits are around accuracy and memory. In the transportation space, the relevant limits are related to attention and reaction time. Whatever industry you design for, spend time clarifying the roles and responsibilities that you've assigned your users and system, and carefully consider if the assignments are appropriately delegated.

Expose the Bias

One of the biggest challenges that analysts face is balancing a variety of different data sources—many of which are incomplete, inaccurate, or, when it comes to some informants, occasionally out-and-out lies.

If an insight is verified through multiple, highly trusted sources, then it should be weighed differently than an insight gleaned from a single suspicious source. As analysts gather information for their reports, it can be difficult for them to keep track of the relative trustworthiness of each insight. A digital interface can help support the analyst by giving visual weight and prominence to more trustworthy sources or well-verified points. A bias check can even be designed, something as simple as a dashboard that lists the number of references in a report and shows how many came from highly trusted versus suspect sources. These sorts of simple checks-and-balances are a great way to reduce the danger of confirmation bias and other intuition-based traps.

This technique can have broader applications across other fields as well, especially in marketplaces where some sources are more trustworthy than others, or on sites where information is crowdsourced. One common example in the consumer space would be showing both the rating and the number of ratings that an item or business has received. (You can see an example of this on Airbnb in Figure 5.8.) By showing the number of ratings, consumers can dial in how much trust they put on the star rating. Most people understand that a 4.7/5-star rating averaged across a thousand people is a significantly stronger endorsement of quality than a 5-star rating from a single reviewer.

Prioritize Ruthlessly

Intelligence analysts have to sift through massive amounts of data, so mitigating information overload is a key goal of the systems that Sam creates. But if you are envisioning Sam's designs as clean, simple screens with a single line graph surrounded by lots of white space, think again. Design techniques around "clearing away the clutter" do not fly with these analysts. The last thing they want is less access to their data. What makes them smile is a single screen with all the right information, all in one place, clearly organized, and ruthlessly prioritized. They want to spend as little time as possible searching for data and as much time as possible thinking about it.

Sam has learned, "You want the system to hold the burden of organizing the complexity of the data. Otherwise, you are giving the user a bunch of data and forcing them to organize it in their brain."

The software that Sam designs doesn't focus on showing less data at once, but rather on organizing massive amounts of data in a way that human minds can spot previously hidden patterns. This might entail taking a single data set and showing it laid out in a dozen different ways. Imagine a screen similar to the one shown in Figure 5.7, filled from edge to edge with graphs looking at the same data from different angles. While it would have taken the analysts days to chart the information by hand, the computer can create the visuals in a matter of seconds. Then it's up to the analysts to spot the chart that shows the critical outlier or the suspicious cluster. That type of insight is something no software on the market can replace yet.

FIGURE 5.7
Although the military interfaces that Sam designs are strictly confidential, the analysis software from Resilient Cognitive Solutions is in a similar vein and uses many of the same principles of design.

Sometimes, hiding away "clutter" *is* the right design choice, especially when designing for nonexperts who are more easily overwhelmed by information overload. But more often than not when designing for experts, *hierarchy* is a more effective design tactic than *hiding*. To make a strong choice about what to include and how to organize it, see the next principle about supporting the decision.

Support the Decision

At any decision point within a flow, designers should ask themselves "What decision are my users trying to make here?" From there, Sam strongly recommends that designers "give the user *all* the information they need to make the decision right in that screen." This is a tenet that can benefit all sorts of decision-focused applications, from ecommerce to dating sites, but it's especially important for expert-level users.

FIGURE 5.8
Airbnb's map view layers the most important pieces of information for users in this stage of rental selection.

Interactive maps like the one in Figure 5.8 from hotel sites like Airbnb do a great job at supporting decisions. The user can simultaneously see the most important decision factors of price and location for spots available on their dates of travel, along with photos and ratings for each location that are just a quick click away.

Supporting the decision doesn't mean that every single piece of available information in the world needs to be on one page. What is often thought of as one decision might actually be a series of many smaller decisions. For a CIA analyst, the smaller decisions might be something like prioritizing what databases to search or deciding on a trust level for each source. For a consumer, before completing a purchase, they often make dozens of small decisions from color selections to shipping methods. Research is often critical to properly determine what is needed for each user decision. *Contextual inquiry*, observing someone while they perform a task, can be especially revealing. UX designer Jorge Zuniga, a former colleague of Sam's, told me when someone uses sticky notes, that's a great indicator that decisions are not being supported properly within an interface. "I have known users to copy a file number from one system onto a sticky note so they could enter it into other systems. That was a big red flag that the system was not properly supporting their work flow."

Protecting Against Bias

Good intuition can be a valuable asset for many types of decisions, but, as discussed in previous chapters, its benefits are limited to situations where there is a reasonably predictable environment and someone has enough time and opportunities to learn what to expect.

For example, what happens when the environment you are designing for isn't predictable or familiar? In some cases, your users' best intentions can actually lead them astray. The first step is simply to identify when the decision point you are designing for is a bias trap.

THE CASE OF...

THE SMOKING SEAMAN

Imagine you are a medic on a U.S. Navy submarine, deep below the ocean surface. A lieutenant comes to you complaining of chest pain. If the seaman is having a heart attack, the protocol says that the ship should surface so the patient can be flown via helicopter to a hospital for emergency surgery. But the electrocardiogram (ECG) and other tests you've run are inconclusive. The ECG came back normal, but the patient's blood pressure was very low. You know that an ECG is not the be-all end-all for diagnosing a heart attack—it could often read as normal, even though a heart attack was imminent. Additionally, you know that this lieutenant smokes three packs of cigarettes a day and he's overweight. Surfacing the ship might jeopardize the current mission, but it would be the right call if it saves this man's life. However, there are any number of nonlethal causes of chest pain. What if the man had just pulled a muscle in his chest? How can you make this call utilizing a rational, reliable decision-making process?

The question from the imaginary medic's story was the very question the Navy hired cardiologist Lee Goldman to answer in the 1970s. For decades prior, doctors around the world had been struggling with diagnosing pending heart attacks. Goldman was one of the first in the medical field to propose using statistical mathematics to identify which patient factors and test results were the best predictors of an imminent cardiac arrest.

By the Numbers

Previous studies had shown that qualities like being male, over-weight, a smoker, or having a high-stress job were linked to a higher risk for heart disease, so the doctors believed that taking note of these markers would help to make better decisions about whom to admit and whom to send home. Goldman's statistical study showed that, although these types of factors certainly showed up more frequently in the population of people who have heart attacks, if a doctor wanted to know if the individual in front of them should be admitted or sent home, these factors were more of a distraction than a help in making that call. According to author Malcolm Gladwell who wrote about Goldman's work in his book *Blink*, Goldman was able to increase the accuracy of his predictions by narrowing down to just a small handful of factors: the results of the ECG, the presence of unstable angina (restricted blood flow), the presence of fluid in the patient's lungs, and the patient's systolic blood pressure.

The use of algorithms in medical decision-making seems old-hat today, but in Goldman's time, many had trouble believing that decisions could get more accurate by looking at less data. It took two decades, and a fairly desperate doctor in an underresourced Chicago hospital, to bring Goldman's work into widespread use. Over a two-year comparison of methods, Dr. Brendan Reilly of the famed Cook County Hospital found that when doctors in his ER used their best judgment based on all the data available to them, they made the correct diagnosis about which patients needed treatment 75–89% of the time. But when the same group of doctors used Goldman's formula with the more limited set of criteria, they got it right 95% of the time. The algorithm was also 70% better at identifying patients who were not having a heart attack, which was especially important for preserving Cook County's very limited resources for patients who truly needed them. With these clear-cut improvements in both costs and patient outcomes, Cook County embraced Goldman's criteria, and the rest of the medical community soon followed.

It takes a lot of rigor to identify bias traps like Goldman did and even more to develop algorithms that can outperform trained professionals. But the formula is only half of what is needed. New tools and procedures must be created so that professionals can incorporate these algorithmic approaches into regular practice, and that's where designers come in.

Implementing New Practices

The first step in implementing a new algorithmic approach meant to replace an intuition-based approach is to clearly define the new decision-making process, breaking it down step by step. Then it's imperative to teach people to use it through hands-on training. The more available, easy, and actionable that users feel the new process is, the less likely they will be to backslide to more familiar ways of working when stress levels rise. The Cook County ER actually hung a poster on its wall detailing the new diagnosis process to ensure that it was highly visible to hospital staff. See Figure 5.9 for an example of a modern-day version of this poster.

Another place where design can really help in implementing new processes is to focus users on the relevant information and wean them off the distracting information so they aren't tempted to continue incorporating it into their decision-making. There is a well-known bias in decision-making that Kahneman calls "what you see is all there is"—meaning that the mind will confidently make decisions based only on the information directly at hand and greatly discount the information that's out of sight, even if those factors are significantly more relevant.

Armed with this information about the human brain, you can actually hijack this bias through visual design to do some good. For example, you should fill the visual field with only the most relevant information for decision-making and hide away the detractors. For instance, in software for HR application tracking systems, designers are experimenting with ways to reduce the influence of unconscious racial and gender bias of hiring managers by hiding information like the candidate's name or image. This process ensures that the candidate's experience and qualifications are the only factors being considered when deciding whom to call back for an interview.

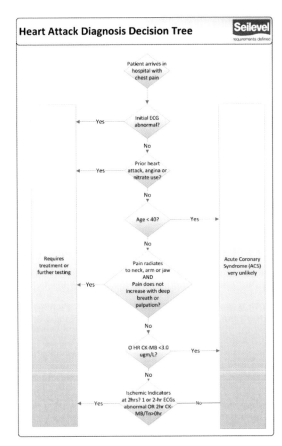

FIGURE 5.9

Similar to what would have hung in Cook County ER in the 1990s, Seilevel's modern-day heart attack diagnosis decision tree utilizes a refined version of Goldman's Criteria.

On the other hand, if it's common within your software for users to be asked to make decisions with incomplete information, put big blanks showing where the missing information should be so that it's clear it's still needed. Keep that information gap top of mind for the user so they can factor it appropriately into both the decision itself and their confidence level.

Calculating Morality

While algorithm-based decision-making is an important tool in fighting false intuition, it's important to understand that these formulas still contain bias. Every formula and algorithm is written by humans, and therefore is influenced by the morals, values, and perspectives of those humans. It's not only impossible to create a bias-free algorithm, but it's counterproductive to try. Instead, creators of these systems should be thoughtful and intentional as to which biases are included, writing algorithms that reflect the morals and values of the communities who will be most affected by them.

During the coronavirus outbreak of 2020, spikes in local outbreaks drove some American healthcare systems past the breaking point, most notably in New York in the early months of the pandemic. Even hardened war reporter W.J. Hennigan found the scale of the tragedy in NYC hospitals hard to comprehend. "When you walk into a major U.S. city and you see the sorts of things that they're dealing with— racks of corpses and industrial warehouses full of people dedicated to processing the dead—it's not something that comes naturally," he reflected during an NPR interview. "The scale of it is incomparable to anything that we've seen." Horrifying images of refrigerated trucks serving as temporary morgues illustrated how close the NYC healthcare system was to total collapse. Day after day, hospital staff were having to make heart-wrenching decisions about how to best ration the overtaxed medical resources of the city—such as which patients got a ventilator, which got a bed in the ICU, and which were sent away to fight the virus isolated and alone at home.

As the nation watched this nightmare scenario unfold, hospitals of all sizes across America began to review their own guidelines for rationing care. These professionals understood that the doctors could not be asked to use their personal judgment in the midst of a crisis. The doctors needed a concrete set of criteria to make these choices. These criteria not only ensured that rationing decisions would be fairer and more equitable, but they also helped hospital staff manage their emotional trauma and guilt of having to make such terrible choices. As discussed in Chapter 4, "Fight, Flight, or Freeze," least-worst decisions are much easier to make when there are clearly defined values. These guidelines were one way to define and codify those values for the medical community. Doctors needed to be able to trust that if they followed the guidelines, they would be adhering to their own values and the values of the communities they served. But how were those guidelines to be defined? What voices should be included in their creation? Whose values should be reflected within?

One of the most prominent sets of guidelines came from Johns Hopkins researchers published in 2019. These guidelines were written through an ambitious, multi-year process involving extensive engagement with members of the general public and the medical community. The researchers worked with a diverse set of community groups to involve members of previously underrepresented communities in the discussion of rationing care.

The research started with a large pre-deliberation survey, which showed that participants shared similar views that care decisions should not factor in race, gender, sexual orientation, religion, socio-economic status, perceived quality of life, ability to pay, or their role in the community.

Fifteen half-day community discussion forums followed the survey, which focused on areas where there was less alignment and asked these questions:

- Is the goal to create a system that will save as many lives as possible or save the most years of life?

- Should priority be given to groups, such as healthcare professionals, so that they can theoretically return to work and save more lives after their own recovery?

- Should priority be given to children? As the report points out, "Prioritizing children over the generation on whom they are directly dependent for their basic welfare raises practical challenges."

- When and how should age play a role?

Of particular concern to many participants were when and how to factor in pre-existing conditions. It is a well-documented reality that historically underresourced communities, specifically Black communities and communities with large numbers of recent immigrants, have reduced access to health care and higher rates of many conditions and diseases, such as obesity, diabetes, and high-blood pressure. To factor in these conditions would be "essentially punishing people for their station in life," as Dr. Lee Daugherty Biddison, the report's lead contributor put it when talking to the *New York Times*.

To address these concerns, the Johns Hopkins model only used a single factor when it came to pre-existing health conditions: Was death likely within a year from these conditions? If yes, it was considered; otherwise, it was not.

The Hopkins model boiled all of these discussions down into a very simple, point-based system considering only three factors:

1. Likelihood of short-term survival

2. Likelihood of long-term survival

3. Age

Like golf, you want to score low on this scale shown in Figure 5.10.

TABLE 1] Proposed Strategy for Ventilator Allocation in Epidemics of Novel Respiratory Pathogens					
		Point System			
Principle	Specification	1	2	3	4
Prognosis for short-term survival	Adults (SOFA) or pediatrics (PELOD-2)	SOFA score ≤ 8 PELOD-2 ≤ 12	SOFA score 9-11 PELOD-2 12-13	SOFA score 12-14 PELOD-2 14-16	SOFA score > 14 PELOD-2 ≥ 17
Prognosis for long-term survival	Prognosis for long-term survival (assessment of comorbid conditions)	...	...	Severe comorbid conditions; death likely within 1 y	...
Secondary consideration					
Lifecycle considerations	Prioritize those who have had the lease chance to live through life's stages (age)	Age 0-49 y	Age 50-69 y	Age 70-84 y	Age ≥ 85 y

FIGURE 5.10

The SOFA score referenced in the chart is the Sequential Organ Failure Assessment. PELOD-2 is the PEdiatric Logistic Organ Dysfunction score. Both scores are based on a set of empirical measurements gathered primarily through blood work, and high scores indicate severe, immediate, life-threatening health issues.

These forms and algorithms guided many difficult decisions during the darkest days of the coronavirus outbreak.

Through extensive community involvement, the Hopkins project attempted to address the idea that no algorithm is truly unbiased. They took the time and did the work to ensure that the bias in the algorithm reflected the values of the community that would be most affected by it.

The Johns Hopkins model was meant to be used primarily in Maryland. When you are creating criteria like this, it is critical to understand that not all communities or subgroups within a community hold the same inherent values. For example, in a similar exercise done in Seattle, their initial survey showed one sharp difference within participant subgroups. "Participants who identified as Hispanic, in particular, felt very strongly about prioritizing children, with 70% indicating that children were 'very much a priority,' as opposed to 27% of non-Hispanics."

As a creator of digital products, you may be asked to take part in the design of choice algorithms or other decision criteria. There are great benefits from involving your user community in creating those algorithms and creating an opportunity for all voices to be heard. Being transparent about the factors involved and how they are weighed can greatly increase the community's trust in the system. Additionally, while gathering community input is important at the beginning of the design process, consider that as the community for your product grows and changes, it may be necessary to re-evaluate your criteria to ensure that it reflects the current values of the full breadth of your users.

As a professional designer, you are paid to create products that reflect the values of your company or client, but your obligations don't end there, especially when you work on products that are used by thousands or even millions of people. Your applications have the ability to allow users to live their values or prevent them from doing so. It's critical to understand the diverse value systems of your users, all of your users. It takes time and work to engage in this kind of research, but it pays dividends in customer loyalty and a more equitable world.

DESIGN QUICK LOOK
MAPPING MORALS

United States is most similar to **Canada**, and most different from **Brunei**
China is most similar to **Thailand**, and most different from **Azerbaijan**

The gray area is the world average.
United States and China are very different

World Ranking (out of 117 Countries)	Preferring Inaction	Sparing Pedestrians	Sparing Females	Sparing the Fit	Sparing the Lawful	Sparing Higher Status	Sparing the Younger	Sparing More	Sparing Humans
United States	35th	67th	47th	37th	95th	48th	49th	14th	68th
China	6th	116th	61st	85th	9th	74th	115th	113th	14th

FIGURE 5.11

In this example screen from MIT's Moral Machine, the graph highlights where cultural norms between the U.S. and China depart.

If you're interested in digging into the way that life-and-death decisions might be weighed differently across different cultures, check out the results of "The Moral Machine Experiment" by Edmond Awad et al., as published in *Nature*. Some of the results of this study are illustrated in Figure 5.11. Click on any two countries on the world map to see how their values compare along many axes such as "Sparing the Young," "Sparing the Higher Status," and "Sparing the Lawful." In the example comparing the values of people in the United States to those in China, the graph shows a strong difference in how those two cultures place value on the young and old. People from China show a strong preference toward sparing the older person in a life-and-death choice, whereas people from the U.S. consistently chose to spare the younger. Elders are highly valued in Chinese culture, especially compared to U.S. culture on average. Preferring to spare the young or spare the old are both types of bias, but biases that are rooted in sacred values. These are the sorts of biases that should not necessarily be stripped from an algorithm-based system, but rather included with intention and balanced to reflect the communities they serve, taking into account the breadth and diversity of values within those communities.

Critical Information: Reasoned Reaction

The best way to help someone behave rationally when under stress is to give them simple, actionable, step-by-step instructions. Unfortunately, sometimes users are forced to deal with high-stress, highly complicated situations. In these situations, users need help from designers more than ever.

We covered four essential questions in this chapter.

How can designers help stressed people who are thrown into the deep end and forced to deal with complexities they have no familiarity with?

- Combine written and visual content.
- Break content up into small, bite-sized pieces.
- Create an interactive experience that allows users to go back and revisit information if needed.

- Use simple, conversational, empathetic language.
- Model behavior that the user should emulate in real life.
- Make the information relatable and help users see themselves in the content.

How can designers help professionals who work in unpredictable environments make the most of their expertise?

- Assign tasks appropriately.
- Expose the bias.
- Prioritize ruthlessly.
- Support the decision.

What types of systems can protect best against harmful bias?

When you are working in a high-complexity environment with many false signals, it is best to create interfaces that help users focus on the factors that will have the best statistical chance of improving outcomes and hiding away information that may seem relevant, but actually has a lower probability of actually improving outcomes. It's also helpful to use visual cues to alert users when they are missing important pieces of information, because the brain tends to discount the importance of any missing information.

Since no designed system can truly be without bias, how can designers ensure that the bias in the system reflects the morals of the communities it is meant to serve?

The only way to ensure that the values of the communities you serve are reflected in the way your product works is to involve voices from those communities in your design process. It takes time, but that time is necessary to increase trust and reduce harmful, unintended consequences, especially for historically underrepresented communities who have traditionally been cut out of the design and decision-making process.

Go to the Source

Neil Armstrong: A Life of Flight: A biography by Jay Barbree, 2014.

"Clinical Trials for Childhood Cancer Drugs Are Critical, but Parents Don't Always Understand What They Are Signing Up For": The original study that inspired the Michigan Medicine work by Laura Sedig and Raymond Hutchinson, 2016.

"Stress and Selective Attention: The Interplay of Mood, Cortisol Levels, and Emotional Information Processing": A study by Mark A. Ellenbogen et al., 2002.

"Prediction of the Need for Intensive Care in Patients Who Come to Emergency Departments with Acute Chest Pain": A study by Lee Goldman et al., 1996.

Blink: A book by Malcolm Gladwell that details the story of Goldman's criteria, 2005.

"The Hardest Questions Doctors May Face: Who Will Be Saved? Who Won't?": A *New York Times* article by Sheri Fink, 2020.

Public Engagement Project on Medical Service Prioritization During an Influenza Pandemic: Guidelines by Public Health: Seattle & King County, 2009.

CHAPTER 6

Recovery

Humans crave mental and physical balance; like Goldilocks, we long to feel "just right." The science-y word for this state is *homeostasis*, and many systems in the body contribute to maintaining it. An acute stress response is meant to be a brief period of extreme activity before the body returns "back" to the normal balance.

A stress response is hard on the body. Important systems like immunity and digestion are suppressed, the heart pounds, the blood flows differently staying closer to your core, and you burn through energy resources quickly. It's not a state you should be in any longer than absolutely necessary.

As a designer, you have the ability to help bring people back down from this heightened state. To do this, it helps to understand the *parasympathetic system*. (*Para* is rooted in the Greek word for "beside," meaning it works alongside the sympathetic system that is in charge of the stress response.) The parasympathetic system is in charge of slowing the stress response, stemming the flow of stress hormones, and reengaging the systems that were put on pause during the crisis moment. If the sympathetic system is all about "fight or flight," the parasympathetic system, illustrated in Figure 6.1, is responsible for "rest and digest" or as some put it "feed and breed."

The parasympathetic system is responsible for many of the visceral reactions that occur in the aftermath of a stressful event, including feelings of exhaustion, triggering of tears, a restart of proper digestion including control of urination and defecation, and even that occasional post-battle uptick in sexual arousal that Hollywood is so fond of using as a romantic plot driver. Engagement of the parasympathetic system is also what signals the body to divert energy toward healing injuries, powering the natural immune system, and rebuilding energy reserves like fat and glucose stores.

As a designer, when you understand the give and take between the sympathetic and parasympathetic systems, you can make the process work for you, helping your user recover faster after a stressful event. In this chapter, we'll explore key questions around calming users through design:

- How long do the effects of a stress response generally last once the parasympathetic system is engaged?
- What aesthetic qualities have a calming effect on the viewer?
- How can designers help users gain closure after a stressful event?
- What is the most essential thing humans need when recovering from a stressful event?

THE PARASYMPATHETIC SYSTEM

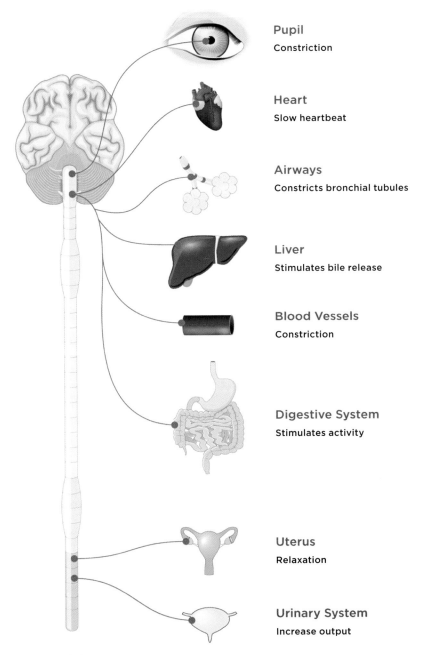

Pupil
Constriction

Heart
Slow heartbeat

Airways
Constricts bronchial tubules

Liver
Stimulates bile release

Blood Vessels
Constriction

Digestive System
Stimulates activity

Uterus
Relaxation

Urinary System
Increase output

FIGURE 6.1

Unlike the sympathetic system, which triggers both conscious and unconscious reactions, the parasympathetic system is largely an unconscious system.

VAGUS, BABY, VAGUS

The vagus nerve is a key player in the balance between the sympathetic and parasympathetic systems. Vagus means wanderer in Latin, and the nerve earned the name by wandering all over your body. It branches from the cerebellum and brain stem and touches nearly every major system in your body, including circulatory, digestive, respiratory, and reproductive. It sends signals from the brain to the body telling these systems to gear up or wind down. It also receives signals from the body that take "gut feelings" back to the brain. You know that feeling you get in your stomach when a situation feels really right or really wrong? That physical sensation is delivered via the vagus nerve. Some call it the highway of the mind-body connection.

The respiratory system is a unique way station along the vagus nerve because it's one of the few parasympathetic systems that you have conscious control over. When you breathe in, the vagus nerve causes a slight increase in your heart rate, and when you breathe out, it causes a slight decrease. If you have a healthy vagal tone, that means you can go from excitement to calm with relative ease. If your vagal tone is poor, a small fright might keep you tweaked for hours.

Normally, you breathe without conscious effort, but if you choose, you can hold your breath or change your breathing rate. This unique feature of the respiratory systems gives you a gateway to "hack" your parasympathetic system via the vagus nerve. Every time you breathe out, the vagus nerve excretes a hormone called acetylcholine. This neurotransmitter is the star of the parasympathetic system and is the juice that prompts your rest-and-digest activities. So deliberately slowing your breath and emphasizing that out breath is essentially a way to self-administer an extra-large dose of acetylcholine with every breath.

Recovery Time

For designers who are trying to predict and shape human behavior, it's helpful for them to understand how long it takes someone to recover after an acute stress response. The exact time, of course, varies significantly based on how intense the scare was, as well as the user's own health, vagal tone, and metabolism. But there are some rules of thumb you can use to estimate, as shown in Figure 6.2.

Estimated Recovery Times

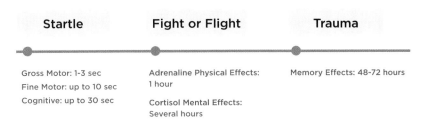

FIGURE 6.2
The recovery time for the effects of stress vary widely, but can be roughly estimated based primarily on the intensity of the response.

Startle Recovery

When someone is startled, their body is momentarily highjacked by their subconscious, which decides which way and how the body should move in order to best avoid the perceived threat. Through Richard Thackray's 1988 summary of collected research on the startle response, he determined that the loss of gross-motor control lasts between 1 and 3 seconds on average, fine-motor skills can take up to 10 seconds to fully recover, and mental effects last the longest of all, with reduced cognitive function lasting up to 30 seconds after a startle. Muriel Woodhead, a Cambridge cognition researcher, pointed out that it was often observed in lab studies that as long as a person's heart rate was elevated after a startle, their cognitive abilities were still impaired. Once their heart rate returned to baseline, so did their mental acuity.

Based on this research, it may be in your user's best interest to give 20–30 seconds of elevated support after a startle while they settle back down. For instance, a driver system may activate select driver assist features directly following the detection of a sudden jerk of the wheel.

Detecting that your user is startled may pose a technical challenge, but there are a wide variety of inputs that can indicate a startle has happened. The skin becomes more conductive when a person is startled, and measuring this conductivity is how most lab studies determine when a participant has been startled. More commonly available inputs from smart devices include:

- A microphone detecting a sudden loud noise
- An accelerometer or motion sensor detecting a sudden jerking movement consistent with a startle
- Increased heart rate from a smart watch or similar biometric tracker

Fight-or-Flight Recovery

The adrenaline and cortisol that are released during a fight-or-flight response are removed from the bloodstream via the kidneys and liver, the same way that drugs and alcohol are filtered from the body. Once the chemicals are removed from the blood, they are then either reabsorbed by the body or eliminated through urination. This is not a particularly fast process, especially compared to the way that huge amounts of stress chemicals can be "dumped" into the blood initially, affecting behavior in microseconds. It takes about an hour for an adrenaline rush to wear off and several hours for cortisol to return to normal levels.

There is one way the body may speed the processing of these hormones: tears. Studies have shown that tears cried after a stressful or emotional event contain high concentrations of stress hormones. While there is still debate about this in the scientific community, these elevated readings have led many scientists to theorize that crying is a way for the body to rid itself of these chemicals more quickly.

Since elevated cortisol has negative effects on memory, learning, creative thinking, and decision-making, designers of systems that rely heavily on any of those skills might consider ways to encourage acutely stressed users to hold off on certain tasks for a few hours until they are in a better frame of mind. Consider offering users a "remind me later" path to relieve concerns that they might forget to return.

If delaying users isn't practical, you can look for ways to provide supplemental guidance during this recovery period. Allstate created an interesting way to provide that additional support to customers who were trying to call for a tow truck after a car accident. Knowing that it is significantly easier for the brain to process information verbally when it is accompanied by a visual aid, they created a feature they called *OmniAssist*, which utilizes both voice and screens that you can see in Figure 6.3.

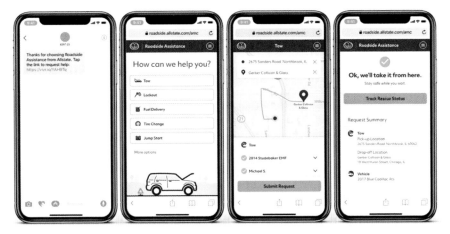

FIGURE 6.3

Allstate's OmniAssist incorporates visual help and automated audio assistance.

The experience starts when a user calls the 1-800 number to request a tow. The automated voice system offers to text users a link where they can complete the process more quickly online. If users agree, the system tells them, "I've just sent you a link. When you receive it, follow the prompts. I'll stay on the line in case there's a problem. Just place your phone on speaker. I'll walk you through making a request." The fully connected system tracks which screens users are looking at in any given moment so the verbal instructions are timed exactly to the step of the process the user is seeing on their screen.

The benefits of this additional support were reflected in a 35% bump in users who were able to complete the order within the automated system and order times that were an average of 3.5 minutes faster. The overwhelmingly positive App Store reviews also spoke to how well it resonated with users.

In a moment after a stressful experience, users often appreciate a bit of extra hand-holding. It's worth getting creative about how to provide that additional assistance in the moments they need it most.

Trauma Recovery

When someone suffers a truly traumatic event, the recovery process has some unique qualities. In most cases, stress will greatly increase the retention of memory around negative events. This is part of our brain's natural *negativity bias* that gives more emotional weight and memory to negative experiences than positive experiences and evolved to help our ancestors remember the many dangers of the wild more accurately. But in rare, extreme situations that involve physical violence, sexual violence, gruesome injury, or witnessing a sudden death, the mind sometimes *disassociates* as a protective mechanism during these horrific events. During disassociation, the mind splits and holds memories separately—a conscious memory and an experiential memory. Disassociation is sometimes described as a feeling of being disconnected from the moment, like watching it happen to someone else. Other times, the mind will actually suppress the memories of the traumatic event.

When memory suppression occurs, it is very rarely permanent. In fact, multiple studies have shown that it's fairly typical for disassociated memories to reintegrate around 48–72 hours after the original event. This timeline means that on the second or third day after a trauma, it's likely for new memories and details to return to the person.

This expected period of memory delay is a critical piece of information for designers of many different types of systems, but especially those in law enforcement. Prior to this research being accepted by law enforcement (and sadly, even still sometimes today), if a victim of a violent crime came forward several days after the traumatic event with new details or information, the police would often accuse them of "changing their story," and their credibility as a witness might be called into question. More modern policing practices now integrate this key understanding of human memory. For instance, the guidelines on handling officer-involved shootings published by the U.S. Department of Justice in 2016 recommended that internal investigators delay personal interviews with other officers who were at the scene "from 48 to 72 hours in order to provide the officer with sufficient recovery time to help enhance recall."

In some individuals, the reintegration of suppressed memories can take even longer—years, if ever. While these traumatic memories may not be consciously recalled, they are still stored in the experiential memory of the hippocampus and can cause dramatic fight-or-flight reactions when the traumatized person is exposed to seemingly innocuous triggers, like a smell or sound they subconsciously associate with their trauma. If you are designing for a population with a high incidence of trauma, such as veterans, refugees, people who are experiencing homelessness, victims of domestic violence, the incarcerated, or first responders, it is essential to proceed with caution and collaborate with experts whenever possible. Even something as simple as a form that assumes a child has two parents can trigger a panic response in a refugee child from a war-torn country who has seen a parent killed in front of them. Check out trauma therapist Dr. Bessel van der Kolk's book *The Body Keeps the Score* for more insight on how traumatized brains work differently.

Calming Aesthetics

The threat-assessment part of the brain is fairly simplistic. The unconscious mind is constantly scanning the environment, looking for things that appear in any way similar to objects that have hurt you in the past. This constant comparison doesn't stop just because you are scanning a webpage instead of a dark forest. Luckily, designing for a calm aesthetic isn't particularly complicated. It's as easy as removing the types of visual elements that might remind a viewer of a dangerous object and increasing the visual elements that look like comfort objects.

Soft and Curvy

In the aptly titled study "Humans Prefer Curved Visual Objects," researchers at the Harvard School of Medicine compared the intuitive reactions of people looking at curved objects versus more angular versions of those same items, as shown in Figure 6.4.

Researchers flashed images in front of participants for 85ms and "subjects were required to make a like/dislike forced-choice decision about each picture based on their immediate, 'gut reaction.'" The responses showed a very consistent preference toward the curved versions of objects. The authors theorized:

> Naturally, a dangerous object (e.g., a knife) can impose a negative sense of threat. However, our results show that a negative bias toward a visual object can be induced not only by the semantic meaning of that object (e.g., "used for cutting"), but also by low-level perceptual properties; even a picture of something as harmless as a watch will be liked less if it has sharp-angled features than if it has curved features.

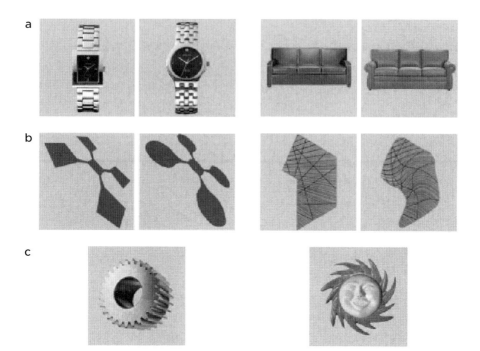

FIGURE 6.4

A figure from "Humans Prefer Curved Visual Objects" that shows the types of images shown to the study's participants.

Calming aesthetics don't need to be sophisticated just because they are speaking to the most primal parts of your user's psyche. It's OK to go with comfort clichés: round corners, soft fades, fuzzy edges, pastel colors, and warm earth tones. Think watercolors, not black ink illustrations. Avoid sharp corners, shouting text, chaotic patterns, and bright, clashing colors. However, remain sensitive to the appropriate aesthetics of the culture(s) you are designing for (remember the lessons learned by BookingLokal discussed in Chapter 3, "Intuitive Assessment"). Deviating too far from the expected look and feel of your brand or industry can trigger feelings of mistrust and destroy your ability to inspire calm through these types of visual changes.

Order and Clarity

One of the most gratifying uses of the internet to date might just be the easy access to "oddly satisfying" images and videos. Click-bait-y titles proclaim, "50 Oddly Satisfying Perfection Photos That Will

Calm You Down. #42 Is Most Satisfying!" These articles are a rabbit hole of perfectly spherical snowballs, tables of precisely patterned color pencils, plants that grow with impeccable petals, and bright shelves of faultlessly placed packages, as shown in Figure 6.5.

FIGURE 6.5
A "perfection photo" from Kickvick.

Humans' innate desire for order is linked to their need for safety. In an environment with a highly regular pattern, anything out of place immediately jumps out and can be identified and dealt with swiftly. In a cluttered, irregular environment, the brain has to work significantly harder to spot approaching dangers. Similarly, white space can evoke feelings of safety because it is the design equivalent of cutting down all the trees for a mile surrounding a castle so that you can easily see your enemy coming.

Your brain is designed to zero in on unknowns and pay attention to them until you can categorize them. When things are partially obscured or unfinished, it leaves you feeling itchy and on edge—like you are missing something and the other shoe may drop. The flip side of that is that there is an innate calmness that comes from being able to see things clearly. Clear things feel safer, more reliable. Kahneman cites several cognitive bias studies where humans perceive information to be more trustworthy when it's easier to read. This example comes from his book *Thinking, Fast and Slow*:

Compare these two statements:

Adolf Hitler was born in 1892.

Adolf Hitler was born in 1887.

Both are false (Hitler was born in 1889), but experiments have shown that the first is more likely to be believed.

Kahneman's research suggests that the use of bold text, as well as high contrast colors between text and background, shorter words, and words that are easier to pronounce, will increase people's intuitive belief that your content and product can be trusted and set their minds at ease.

Natural Elements

However, as much as precision and clarity invoke calm, the everything-in-its-place aesthetic taken to the extreme robs you of something. It can become sterile, restrictive, and unnatural. Humans are imperfect creatures, and they grate against expectations of ridged perfection. Design elements that incorporate natural irregularities have their own calming power.

In some ways, it seems illogical. From what we've discussed so far, uncertainty should be stressful, not soothing. But when the inconsistencies are in the form of nonthreatening natural sources like a babbling brook, or the sound of wind chimes, and fall within a normal range of safe, expected behavior, they are quite calming. No doubt these instinctive preferences are remnants of thousands of generations of humans who evolved in the pre-industrialization era.

Biophilic design, long popular in Eastern architecture and now gaining popularity in Western architecture and interior design as well, recognizes the inherent draw of nature and its profound impact on human well-being. See Figure 6.6 for an architectural example. Having access to nature is known to reduce stress and improve health.

FIGURE 6.6
A classroom that models biophilic design principles from Axis House.

One of the greatest gifts that Florence Nightingale's work in the 19th century gave to hospital patients in the 20th century was to popularize the idea that fresh air and sunlight were critical components of healing. Nightingale, famous for her work as a nurse, was also a talented statistician and gathered extensive data showing the link between health and access to nature. Her findings have been consistently supported through more modern research. For instance, in a 1984 study published in *Science* magazine, patients placed in rooms with windows were shown to recover faster from surgery than patients placed in windowless rooms. Additional studies performed over the last half-century have tied the incorporation of nature and natural elements in interior design to better physical health, mental health, cognitive function, and creativity. It is largely accepted by professionals in workplace design that the incorporation of natural elements, from sunlight to plants to water features, is a critical part of designing a healthy work environment.

Visual Designs

Use:

- Rounded corners
- Effects that look soft or fuzzy
- Images of nature
- Images of people smiling sincere smiles or with other sincere, friendly expressions
- Regular patterns, everything in boxes
- Symmetry

Avoid:

- Sharp, spiky edges
- Busy, conflicting patterns
- Eyes looking directly at the user, especially predator animals, including humans
- Images of people looking directly at the camera with angry or even neutral expressions (neutral can be misread as aggressive during stressful moments)
- Images of people with fake smiles

If you primarily design digital products, you may be wondering how these studies on the calming powers of nature are relevant to you, but the natural world is a rich source of design inspiration for experiences that seek to evoke more peaceful emotions. In the spirit of *biomimicry* (modeling man-made objects after natural objects and processes), you can look to calm movements like flowing water or light breezes to inspire interface animations. Consider serene outdoor environments like mountain ponds or pastures for color palettes. Use creative design prompts like, "What would this design look like if it grew naturally from the seed of the user's intent?" If nothing else, consider ways to incorporate pictures of nature into your design—still images of nature aren't quite as effective at eliciting calm as views of actual nature, but they do still have measurable effects. Humans' connection to nature is so strong that they can draw comfort from it even via pixels.

Sounds

Use:

- Slow, irregular sounds that mimic rain, waves, or wind
- Tones in major intervals and harmony

Avoid:

- High-pitched screeching noises similar to screams
- Low rumbly noises
- Dissonant tones that do not resolve to harmony
- Tones that rise and rise in pitch (sometimes called Shepard tones)

Motion Design

Use:

- Slow irregular movements that mimic natural motion, such as leaves blowing in the wind or light reflecting off slowly moving water
- Rhythms similar to slow breathing

Avoid:

- Rapid movements or blinking
- Animations rushing rapidly toward the viewer
- Excessive bouncing

Gaining Closure After a Stressful Event

You've learned how the acute stress response is supposed to work: something stressful happens, stress-response hormones are dumped into the bloodstream, the resulting burst of energy is used to deal with the threat and reach safety, and then the parasympathetic system pops on to take the body back to homeostasis. But what happens when the person experiencing the stress response is unable to take action?

In his book *The Body Keeps the Score*, trauma therapist Bessel van der Kolk compares the experiences of being allowed to take *effective action*— an action "that ends the threat"—and *immobilization*, which "keeps the body in a state of inescapable shock and learned helplessness." He illustrates his concept using the photos in Figure 6.7.

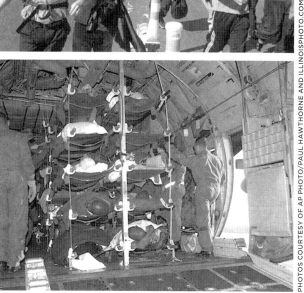

PHOTOS COURTESY OF AP PHOTO/PAUL HAWTHORNE AND ILLINOISPHOTO.COM

FIGURE 6.7
Images of people running from Ground Zero on 9/11 (top) and survivors of Hurricane Katrina being evacuated (bottom) from *The Body Keeps the Score*.

The image on the top shows New Yorkers running from the site of the Twin Towers as they collapsed on 9/11. When counseling survivors of 9/11, van der Kolk found that those who had the experience of running away from the site and reaching a safe location had surprisingly fast recoveries from the trauma they experienced.

"Faced with danger, people automatically secrete stress hormones to fuel resistance and escape. The brain and body are programmed to run for home, where safety can be restored and stress hormones can come to rest." Reaching safety closes the loop on the stress response and gives the parasympathetic system the cue to bring stress levels back down.

In contrast is the experience of the patients in the photo on the bottom who were too ill or injured to be transported while seated when being evacuated after Hurricane Katrina. "In these strapped down men who are being evacuated far from home... stress hormone levels remain elevated," van der Kolk explained, "stimulating ongoing fear, depression, rage, and physical disease."

Humans are not meant to sit still during a crisis. Even if they aren't in a full fight-or-flight response, stress drives activity. It can lead to seemingly illogical or odd human behavior. Consider the baking craze that cleared out the flour supplies in grocery stores across America during the first few months of the coronavirus outbreak in 2020. The physical act of baking was subconsciously used as a therapeutic release by millions of people trapped in their homes by shelter-in-place orders. When seen through the primal lens of a survival instinct, making fresh-baked provisions high in sugar and fat was a fitting use of the nervous energy caused by daily news stories about dwindling food supplies.

Digital designers are well practiced in helping users take "effective action." A core part of your job as a designer is to ensure that users "do the thing," whatever "the thing" is for your application. One of the keys to maximizing user action is to understand that the moment of peak emotion (which is often also the moment of peak stress) in any experience is the moment when users are most motivated to act. It's imperative to ensure that the ability to take an effective action is most easily accessible right in that moment.

This cycle of stress, drive, action, and recovery is critical for designers to understand. The success of your products often depends on predicting where and when users will feel that peak of stress that spurs them to action and allowing users to take action exactly in that moment. When timed correctly, that action gives users a satisfying sense of closure, jump-starting their parasympathetic system and their return to homeostasis.

I worked on an exhibit in 2015 called "The Unforgotten." This was a powerful memorial for victims of gun violence created by FCB Chicago in partnership with the Illinois Council on Handgun Violence. An image from the exhibition is shown in Figure 6.8.

PHOTO COURTESY OF FCB CHICAGO

FIGURE 6.8
Visitors viewed statues dressed in the clothes of the victims of gun violence at "The Unforgotten."

The haunting, faceless statues, made from the clothes of the victims of gun violence, each wore a name tag, introducing the individuals who had been killed. Those people who attended the exhibition were

The Most Essential Thing: Human Connection

During the Blitz of World War II, as the Germans rained down bombs on London, many sent their children to the country to stay with friends and family farther away from the horrors of war. Decades later, as researchers began to study the effect of war on children, it was discovered that the children who were separated from their families had higher rates of lasting emotional trauma than the children who witnessed bombings, destruction, and death while remaining with their family members in the city. Human contact in a time of stress makes people mentally stronger, and companionship is

prompted to download an app on their phones, which, when pointed at a name tag, would play videos about that person, edited with interviews from family members, urging the viewer to take action to stop gun violence (see Figure 6.9).

PHOTO COURTESY OF FCB CHICAGO

FIGURE 6.9
The app for the exhibit could recognize the name tag on the statue and bring up videos from that person's life.

In this moment, just after the peak of the emotional appeal, the app gave viewers the ability to add their signature to a petition urging politicians to change state and national laws around guns. This changed viewers into activists and emotional stress into effective action. This action not only served the cause of the exhibition, but it also helped the viewers process their difficult emotions more productively.

a critical part of how humans process stress. Therapists often emphasize the critical importance of finding the right fit between therapist and patient. When that human connection clicks, it is often as effective as medication in helping people improve their mental health.

People need people, especially when they are recovering from a trying experience. As a designer of human-to-computer interactions, you might find it easy to lose sight of your role in that human-to-human connection, but there are many ways that your designs can contribute to this essential bond. Sometimes your product will connect people directly, sometimes it will help them perform a task more quickly so they can get back to the people in their lives, and sometimes design can help break through barriers between people, allowing healing to begin.

THE CLOSE CALL

Alex Durussel-Baker was in a plane taxiing out of the Edinburgh airport listening to a flight attendant explain safety procedures when she saw her doctor's office was calling. She'd had blood drawn for a few tests just hours earlier, and though she technically should have shut her phone down for take-off, something told her she needed to answer that call. It was a good thing she did. Ducking down low in her seat, she listened nervously as her doctor told her bluntly that her blood work showed such alarmingly high glucose rates, she was just a step away from falling into a diabetic coma. "You need to go directly to hospital the moment you land," her physician insisted.

Alex followed her doctor's instructions and spent the next three days in a New York City hospital, alone in a foreign country, completely overwhelmed by what they were telling her: she had type 1 diabetes, an incurable condition that would require a lifetime of finger pricks, insulin injections, careful carb counting, and constant vigilance.

When Alex finally got home to Edinburgh, she continued to struggle with information overload related to her diagnosis. She now wore a digital glucose monitor and checked her blood sugar dozens of times a day. "When you first start, your life just becomes regimented by numbers. It's mad," Alex said. If her levels were even a touch off the recommended range, the app she used would flash orange or red screens at her, constantly making her feel she was doing something wrong. Managing her levels became an all-consuming activity. "It was a bit like OCD. I was like, 'I need perfect numbers and nothing else will do.'"

As engrossing as this disease was for Alex, it was equally bewildering to her family. Although they tried to be supportive, they didn't really know what to say to her. "My dad and sister especially did not know how to take this illness so they would just never talk about it. For me it was a 24/7 thing, I was checking my blood every 15 minutes. It became this elephant in the room." Alex felt isolated from some of her most important support figures at an extremely vulnerable time in her life.

Professionally, Alex is a product designer, and she works with clients to break down complex problems and create workable solutions. She decided to apply some of her professional methodologies to the convoluted and isolating reality of living with type 1 diabetes. "This disease is so complex, touches on so many elements of your life, and there are so many misunderstandings." Alex realized, in order to process everything herself and also allow her family to better connect to what she was going through, she needed to do two things: make the invisible tangible and break down the complexity. She decided to take one element a day and make a poster about it.

By designing these bold, beautiful posters (shown in Figures 6.10 and 6.11), Alex could focus on one piece at a time.

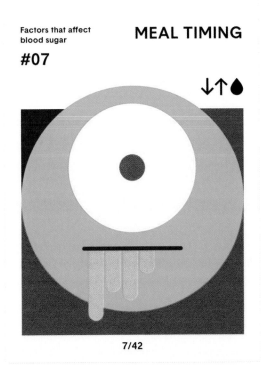

Factors that affect blood sugar

#07

MEAL TIMING

↓↑⬤

7/42

FIGURE 6.10
The posters from Diabetes by Design make education more accessible by adding a sense of whimsy to an otherwise heavy topic.

She also created humorous posters that spoke to various elements of her new reality. And she began posting all of her posters on Instagram along with short essays.

FIGURE 6.11
Alex included short essays for each poster on her Instagram account, which gave additional context and perspective to her followers.

Soon the Instagram feed became an in-person exhibition. Later, Alex was invited to take part in the UK's largest conference on diabetes, and doctors began approaching her about using her work to help patients process their new normal. They loved the way her work boiled down the complexities of type 1 diabetes and made it understandable and consumable. In 2021, Alex gathered her sizable collection of art, worked extensively with diabetes medical specialists, tested with other diabetic patients, and finally launched a complete set of "T1D Companion Cards," shown in Figure 6.12, that are equal parts teaching tools, entertainment, and conversation starters for the topic of type 1 diabetes.

Beyond the impact on diabetes education, Alex understands that her work has an important role to play in opening the door to human relationships. "If you break an arm, and you're walking around with a splint, people will say, 'Oh, what happened?' And you've got a cool story or it makes a nice ice breaker. These posters became my broken arm and what people came to talk to me about."

Having her art as a way to open a discussion made a huge difference in Alex's relationship with her family. "It became something we could connect over. It was empowering because suddenly I had the ability to educate them and share something about what

I'm experiencing." Fans of Alex's work find it has opened doors to conversations with their own family members

Yes, I can eat that.

MYTHBUSTING:
No such thing as a diabetic diet

For a very long time, people with Type 1 diabetes were given very strict diets and insulin regimens to follow in order to keep their blood sugar from going dangerously high or low.

In the last 10 years, thanks to modern insulins and technology like continuous or flash glucose monitors, people with type 1 diabetes can now eat what they want and even splurge on occasion - just as much as a someone without diabetes!

However, a very carb or fat heavy meal takes more diabetes math (more on that in the *Riding the sugar wave* set) but basically if we can carb count it, we can eat it.

So yes, I can eat that donut, now that you mention it, I might just have two. So step aside, I'm packing needles.

Scan to hear more on this topic

FIGURE 6.12
T1D Companion Cards are used to educate patients and serve as conversation starters for discussing diabetes with friends and family.

Enabling human connection is the most powerful thing a design can do to help someone recover after a stressful experience. However, more often than not, UX designers are asked to create digital interfaces that replace face-to-face interactions. Even when practicing human-centered design, often designers focus in narrowly on the direct user and that person's interactions with their product, forgetting that the user is also part of a family, community, and society. Try to avoid falling into this trap. Widen your perspective to consider how your product can enable your user to partake more fully in these wider human relationships. Research, such as user interviews, can help you map out those wider circles of relationships. From there, look for opportunities to help your user share their personal experiences with others in a way that gets them the support and connection they need.

Critical Information: Recovery

While the acute stress response has short-term benefits for survival, it is a state that suppresses many vital physical functions, so people should not remain in this state for any longer than absolutely necessary. Designers should help their users regain homeostasis, a baseline state of balance, as quickly as possible after a stressful event.

How long do the effects of a stress response generally last once the parasympathetic system is engaged?

The amount of time it takes the parasympathetic system to counteract a stress response depends on how severe the reaction is. A simple startle reflex can take up to 30 seconds to wear off. A fight-or-flight response can have adrenaline effects that last about an hour and cortisol effects that last for several hours. A traumatic experience often affects memories for 48–72 hours, but can have lasting mental and physical effects that last years or even a lifetime.

What aesthetic qualities have a calming effect on the viewer?

When trying to create experiences that elicit a sense of calm in the user, use visual cues from objects that signal safety—for example, objects that are soft, curvy, clean, orderly, and plentiful. Ensure that nothing seems hidden or obscured. Also look to nature for inspiration, incorporating elements such as earth-tone colors, gentle irregular movements that mimic rippling water or leaves waving in a breeze, and objects, images, or textures from natural environments.

How can designers help users gain closure after a stressful event?

Allowing users to take an effective action during a peak moment of stress is one of the best ways that a designer can help their users expel the built-up energy from stress hormones. Completing an action or reaching a safe space is often the trigger that kicks the parasympathetic system into gear. Forcing someone into a state of immobilization during a stress response keeps them in an elevated state, which can have lasting negative mental and physical health effects. Designers should strive to place a path to action as close as possible to the peak moment of stress in any experience.

What is the most essential thing humans need when recovering from a stressful event?

Human connection is the most effective calming tool available. Even if your product is normally meant to replace face-to-face interactions, consider ways that it might still enable human connection for users who have recently undergone a stressful experience.

Go to the Source

"The Neurobiology of Grace Under Pressure": An article about the role of the vagus nerve from *Psychology Today*, by Christopher Bergland, 2013.

"Performing a Visual Task in the Vicinity of Reproduced Sonic Bangs": A study on the startle reflex by Muriel M. Woodhead, 1969.

The Body Keeps the Score: A book on trauma by Bessel van der Kolk M.D., 2014.

"Humans Prefer Curved Visual Objects": A study from *Psychological Science* by Moshe Bar and Maital Neta, 2005.

"50 Oddly Satisfying Perfection Photos That Will Calm You Down. #42 Is Most Satisfying!": An article on kickvick.com.

"Florence Nightingale, Datajournalist: Information Has Always Been Beautiful": An article from *The Guardian* by Simon Rogers, 2010.

"What World War II's 'Operation Pied Piper' Taught Us About the Trauma of Family Separations": An article in the *Washington Post* by Amy B Wang, 2018.

CHAPTER 7

Alarms and Alerts

You've now learned about designing for each of the five stages of the stress response, but what about those moments when someone isn't yet aware of a pending crisis? What is the best way for a designed system to tell a human user that something is wrong? *Alerts* draw a user's attention to something and *alarms* draw a user's attention to something dangerous. Both are essential components of contingency design. They allow humans to intervene or escape in a crisis situation. Alarms and alerts can take many forms: auditory, visual, or haptic (vibration), but often the most effective alert designs use a combination of all these mediums.

Loud noises are often the "go-to" for alarms and for good reason. For people without auditory impairments, sound elicits the fastest reaction time of all of the senses. It is also omnidirectional, a critical feature that makes it ideal for receiving unexpected information because you don't have to be facing a certain direction to receive the signal. A typical field of view for a human is limited to about 200 degrees left to right and 135 degrees up and down. But you can easily and immediately hear noises that happen anywhere within your vicinity, even directly behind you. Light-based alarms have to throw bright light around, bouncing it off surfaces, to try to get the signal into the much more limited field of vision and ensure that it catches the viewer's attention.

But auditory alarms and alerts are so effective, and so cheap and easy to add to electronic devices, they've become overused, overrunning the soundscape of the modern era. They are often the first sound that people hear in the morning from their bedside tables, and they continue in a constant barrage during the day, ringing, binging, beeping, blaring, and interrupting. The toaster, the coffee machine, trucks backing up, each email received or sent, each IM, call, text, tweet, or notification is yet another ping in a constant stream of noise and distraction.

NOTE INCLUSIVE ALARMS

No matter how effective sound is at capturing attention, always be sure to include nonauditory components to the alarms and alerts you design. According to the National Institute on Deafness and Other Communication Disorders, roughly 2% of adults age 45–55 have disabling hearing loss. That number rises to 50% for those who are 75 and older. Those numbers don't reflect the even larger number of users who may be temporarily unable to hear your alerts due to circumstances, such as wearing earphones or being in a loud environment.

Dr. Judy Edworthy, dubbed the "Godmother of Alarms" by the *New York Times*, is a professor of applied science at Plymouth University in Britain. She says an average person hears nearly 100 alarms and alerts per day. And this endless noise is taking a toll on people's health. According to a study published by the World Health Organization, in Western Europe alone, at least a million healthy life years are lost annually from stress caused by excess noise.

And ironically, it's hospitals where the soundscape has gotten especially dangerous in recent decades. An investigation by the *Boston Globe* revealed alarm-related mistakes were linked to more than 200 deaths in the U.S. between 2005 and 2010. At the root of the issue was the sheer number of alarms that go off inside hospitals. One study from the University of California San Francisco logged an average of 187 alarms per bed per day. A similar study from Johns Hopkins Hospital found that staff heard a critical-level alarm every 90 seconds. Exacerbating the problem is the shockingly high rate of false positives—a PubMed study from 2013 found 72% to 99% of alarms in clinical settings are actually false alarms.

Unsurprisingly, the frequency and inaccuracy of these alarms leads to problems in the performance of medical staff. It's a state referred to as *alarm fatigue*, and it can have deadly consequences for patients. Sometimes, the problem comes when alarms are unnoticed or ignored, but just as often issues are caused by hospital staff breaking protocol to change settings or turn the alarms off entirely because the noises are so disruptive to both staff and patients.

In addition to the frequency of false alarms, the alarms themselves aren't particularly helpful. Most medical alarms rely on electronic beeps to convey information. Different tunes and rhythms are supposed to convey different messages. The problem is, humans have a fairly limited ability to distinguish between these types of signals, especially when the tones themselves are all variations of the same "beep" sound. At maximum, they can learn and distinguish about 5–8 patterns. However some machines produce an absurd 50 or more alarm variations, all with supposedly different meanings. And that's just a single machine among a dozen or more in an OR. Getting any inherent meaning from these alarms becomes almost impossible for hospital staff, and yet device manufacturers continue to design their alarms as though every doctor and nurse is as fluent in robot-speak as Luke Skywalker conversing with R2D2.

In the mid-2010s, the medical community really began to take the issue of alarm design seriously. Many healthcare organizations worldwide set their sights on addressing "alarm hazards," including The Joint Commission (www.jointcommission.org), which named addressing alarm fatigue a top U.S. national patient safety goal in 2014. It was clear that new global standards would be needed to allow a cohesive alarm strategy across products from various manufacturers. The new standards would need solid, evidence-backed recommendations for providing safe, meaningful alerts and reducing harmful, unnecessary noise. However, as Edworthy put it, "Changing and updating standards is akin to the proverbial changing of the course of a ship using a teaspoon."

To increase the size of her teaspoon, Edworthy joined forces with a dream team of sound designers and scientists from around the globe, including Dr. Elif Ozcan, who leads the Critical Alarms Lab at the Delft University of Technology in the Netherlands; Yoko Sen, an electronic musician and sound designer; and Dr. Joseph Schlesinger, a jazz musician and anesthesiologist at Vanderbilt Medical Center; among others.

While you may or may not design medical devices, the research and ideas that have come out of their work can give insights and inspiration to designs involving all sorts of alarms and alerts. In this chapter, we'll look at some of the more innovative solutions being proposed by Edworthy's team and others while we explore these key questions:

- How can designers create systems that communicate status changes calmly yet clearly?
- What techniques can designers employ to make alarms and alerts smarter and more context aware?
- How can designers maximize the meaning intuitively conveyed through each alarm signal?

Creating Systems to Communicate Status Changes

When she was a young adult, electronic music artist Yoko Sen underwent treatment for a life-threatening disease that required her to spend many nights in hospitals. As a musician, Sen was especially aware of all the sounds in her environment. There was one machine in her room that would periodically beep. When Sen asked her nurse if she should be concerned by the noises the machine was making, her nurse told her to ignore it. "That machine just beeps."

In her book *Calm Technology*, Amber Case makes a case that to truly serve human needs in a way that creates a calm environment, digital products should require the smallest amount of attention possible to allow users to achieve their goals. In a hospital environment, there is the goal of understanding the status of a patient's vital signs and being alerted by any changes, but the constant beeping Sen had to listen to was not an appropriate match for the amount of attention that needed to be paid to that machine on an ongoing basis.

Humans tend to worry when they don't know the status of what's happening, especially when they are waiting for something important that is out of their control. One thing that technology does much better than humans is to monitor boring things. So technology is often used to do long-term monitoring and alert people when something important happens. But batteries run out, Wi-Fi goes down, and "stuff" happens. Especially when you are waiting for something important, you need a way to answer questions like, "Is this thing still on?" Case says the best designs "answer such questions by giving you information in an unobtrusive way."

In terms of design solutions, Case writes that, "A status light is perhaps the calmest way of conveying a piece of information." The communication is small, unobtrusive, and passive, often sitting in the periphery, waiting patiently for the user to glance its way. Think of the light on a surge protector. It doesn't require a click or query to know that it's getting power. The information is easy to find when you want it, easy to ignore when you don't. But, of course, this isn't the only way to solve the problem of status updates.

For many physical products, you don't always even need a digital solution to give a status. Whether it's fluids in an IV bag or food in a self-refilling dog dish—if the container itself is see-through, you can tell at a glance how close it is to being empty. This kind of quick status check isn't precise—you can't know for certain that the dog food container is exactly 83.1% empty, but you can tell at a glance that it's time to grab a new bag of dog food as you're headed out the door to the grocery store.

Not all information needs to be delivered at the same fidelity, or level of precision and specificity. One way to ensure that information can be understood with a short glance is to lower the fidelity of the information presented. Give your users only as much information as they absolutely need, especially at the initial level of contact or display. When you have a large amount of data you'd like to convey at low fidelity, that can actually be a really fun place for designers to play, because there are so many ways to convey data and trends in data when you move away from a literal display of numbers.

Hearing the Data

At Critical Alarms Lab, Dr. Elif Ozcan has spearheaded several projects that look at hospital data in new ways. They've tackled everything from animated artwork that moves and changes colors based on the sound levels of the surrounding hospital room, to a sound-based project called *CareTunes*, which replaces the competing, beeping monitors with a more harmonious and usable soundscape that lets hospital staff hear changes in data.

For CareTunes, Ozcan brought in Yoko Sen, who had ignited the imagination of TEDMED (the Medical arm of the famous TED Talk programs) with her talk about the last sound people want to hear before they die (spoiler alert: it's not a medical alarm.)

In her talk, Sen shared how the constantly beeping medical machines that surrounded her during her hospital stay created a cacophony that, for a trained musician, was a special type of torture. "I remember some monitor kept beeping, and it was the note of C. And across the hall something was beeping at a high-pitched F sharp." That jump from C to F sharp is called a diminished 5th by musicians. "It's a type of dissonance that people for centuries called 'devil's chord.'"

For those of you who may have forgotten (or never learned) your music theory, let's review a few common musical concepts and the feelings they evoke. You've likely heard the terms *major key* and *minor key*. Songs written in major keys often sound happy (think of many children's songs such as "Twinkle Twinkle Little Star" or "Mary Had a Little Lamb"). Songs written in minor keys sound sad, wistful, or even scary—think Beethoven's "5th Symphony," "Hotel California" by the Eagles, or Rihanna's "Desperado."

Dissonance in music happens when two notes that clash are played together. An example of this would be the chord played as the stabbing noises in the infamous *Psycho* movie score. Musicians often use dissonance more subtly in music to create tension, which is then *resolved* when the chord switches to a *harmonic* chord. That resolution feels good, like scratching an itch. For an example of this, think of the opening three notes of *The Simpson's* theme song. When the singers sing "The Simp-sons" at the beginning of the song, the jump from the first note ("the") to the second ("Simp-") is a diminished 5th, that "devil's chord" Sen referred to earlier. Then the third note ("-sons") is a major 5th from the original note, which resolves the dissonance and gives the feeling of the clouds parting and the sun shining through, which is exactly what happens on-screen.

But in Sen's hospital room, that dissonance caused by the various monitors never was resolved. It just kept scraping against her ears like nails on a chalkboard. She found the soundscape of her hospital room to be a "permanent state of chaos ... as if to amplify my internal sense of fear, confusion, and loss of agency." As Sen lay in bed, she kept asking herself one question that would ultimately lead her toward a major career shift, "Can't I just tune these beeps?"

Years later, she would get a chance to do just that through the Critical Alarms Lab's conceptual prototype, CareTunes. Sen, Ozcan, and their colleagues turned patient data into calming background music that could draw attention without causing undo alarm when the data changed. The patient's heartbeat became the base beat of the song, their oxygen level was represented by a guitar chord every couple of measures, and the blood pressure generated a rising or falling melody. If one of these data points began to rise, the notes of that instrument rose along with it. If the vital sign began to creep outside the appropriate range, the instrument played a minor chord instead of a major chord, signaling that something was wrong and needed attention. If things continued to worsen, the instrument took on the "womp womp" tenor of a sad trombone noise. The more dangerous the level, the more obvious and attention-grabbing the dissonance and the greater sense of unease that was generated in the listener. An accompanying visual display, shown in Figure 7.1, provided more detailed information when needed.

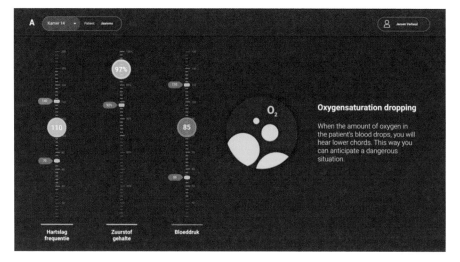

FIGURE 7.1

The CareTunes concept from Critical Alarms Lab uses a well-tuned symphony to communicate patient status, accompanied by a screen showing critical metrics such as heart rate, oxygen levels, and blood pressure.

While this concept would need more testing and development to ensure that it is fully understood by people from different cultures with diverse musical traditions, as well as by people who are hearing impaired or tone deaf, this design does exactly what Sen had longed to do while lying in that hospital bed years before: it tunes all of the various device sounds into a single, harmonious symphony. A symphony that could be far more intuitive than machines that produce dissonant beeps even when everything is going well. A symphony that allows doctors and patients to be reassured by the creative application of low-fidelity data that things are going well, through a pleasant, calming tune that can fade to the background of consciousness until it is needed.

Mode Switching

While the medical field is trying to reduce the strength of the signals sent about minor status changes, the transportation industry struggles with a related issue, but on the opposite end of the spectrum: how to make it clearer when automatic pilot systems shut off partially or completely and pass control back to the user. Many lives have been lost in plane crashes caused by *mode confusion*, when the human pilot either doesn't understand or forgets which level of automation is engaged and is unclear which tasks are their responsibility and which the system will perform automatically. What started as a problem unique to aviation is spreading across the transportation sector as more and more trucks, buses, trains, and consumer vehicles incorporate partial or full levels of automation.

This issue of mode confusion is one that the aviation industry has been working to address for decades—primarily through pilot training, although there have been some attempts to improve displays and interfaces to make it more apparent which mode is active at any given time and to signal when a change has occurred. When it comes to consumer vehicles, so far car manufacturers have been left to their own devices to figure out how to convey mode changes to drivers and have created a wide variety of experiences that do so with varying degrees of failure.

Missy Cummings is a professor of Electrical and Computer Engineering at Duke University who studies the effects of mode confusion in self-driving cars. Prior to her career in academia, Cummings was one of the Navy's first female fighter pilots. She was well-versed in issues of mode confusion from her pilot training. She even had some friends in the Navy who had some close calls due to it. She recalled one incident when one of her peers forgot to put his weapons back into safe mode after returning to the aircraft carrier.

"Right before he got back [to the carrier], his commanding officer in the other plane decided that they would do a fun one-v.-one, which was like a dogfight," Cummings shared in an interview with *Electronic Engineering Times*. "My friend, I like to call him Spider (not his real call sign), got the jump on the commanding officer and got in position to fire. There's this really compelling shoot queue. So, the system will scream at you to shoot. But you needed to make sure the letters 'SIM' were beneath, to show you were in simulated mode. Right? But he didn't double-check." The font in the display, Cummings explained, was so small that it was quite easy to miss the fact that it wasn't showing. "He pulled the trigger, thinking he was in simulated mode." The plane launched a live missile, headed directly toward the commanding officer's plane. Thankfully for all involved, the shot missed, skimming just below the officer's aircraft, but it was an extremely close call.

This story perfectly illustrates how even elite pilots with extensive training can get confused about what mode they are in. In order to avoid mode confusion among general consumers, mode switching alerts and mode statuses must be obvious and intuitive, especially given the fact that most drivers have *no* training on automated systems. Cummings and her students have done testing on auto-mated driving systems across several consumer car manufacturers and found that all of the systems tested had major issues with mode confusion. Her major concern coming out of these tests was the amount of inconsistency, even in how a single vehicle behaved. "No car performed the same way two times in the road on any test."

When asked how she might improve mode awareness in average drivers, Cummings joked, "If I were the queen of the world, every time [Tesla] Autopilot or any system like Autopilot turned itself off, I'd have a big red siren in the car going 'whoo, whoo.'" On a more serious note, she did point to Cadillac's Super Cruise as a best-in-class example of communicating mode switches and statuses to a driver, which is shown in Figure 7.2.

FIGURE 7.2

The large, bright status light for Cadillac's Super Cruise is placed along the top of the steering wheel, well within the driver's visual field.

Cadillac's Super Cruise system allows drivers to take their hands off the wheel and feet off the pedals in certain well-controlled driving conditions like highways. When a driver first engages the system, a green bar of light appears at the top of the steering wheel and stays on as long as the system is piloting the vehicle. The green bar turns red and flashes to alert the driver if the Super Cruise system has to disengage for a technical reason, like an obstructed sensor.

As an additional safety precaution, drivers are required to keep their eyes on the road. To enforce this, in-vehicle sensors track the driver's attention, and if they detect a driver's gaze is straying for too long, the system first flashes a red bar on the steering wheel. If the driver continues to ignore the road, a voice will prompt the driver to pay attention. And finally, if the driver refuses to end their game of Candy Crush despite these copious warnings, the car will actually pull itself to the side of the road and come to a stop.

A strong step forward in addressing the growing issue of mode confusion in self-driving vehicles is the placement, size, and prominence of the status light. The hope is that other manufacturers will follow suit with similar designs, although additional considerations and testing may be warranted to ensure that the system is fully accessible for those who are color blind.

Alarms That Are Smarter and More Aware

THE SHOUTING SURGEON

The surgery was nearly done, and Dr. Joseph Schlesinger, the anesthesiologist, was monitoring the patient during the final steps, preparing to wake the patient from anesthesia-induced sleep. Music was playing loudly in the OR. It had been a long, grueling surgery, but the task had been made more tolerable by some great music playing in the background throughout. (Playing music during surgery is now a fairly common practice since it has been shown to improve surgical efficiency.) But then, in celebration of a successful procedure, the surgeon decided to crank up the volume of the music. About that time, things started going wrong.

"Now my alarms are going off, the music volume is high. The surgeon and I are having to raise our voices and almost yell at each other to communicate, not because we're mad at each other but just to understand each other."

Schlesinger recounted this case study story in an episode of the design podcast "99% Invisible," and said this experience gave him an idea. He and another anesthesiologist began collaborating on a device they call the *Canary Box* that would connect the alarm system and the music system and allow the various auditory outputs in the OR to work together. Schlesinger explained, "When the alarm is in the warning zone, it halves the music volume. And when the alarm is in the danger zone, it turns the music off."

This idea speaks to a greater human need when it comes to alerts and alarms—for the technology that surrounds people to be more contextually aware when trying to communicate with them. Most people expect a human companion to "read the room" and adjust volume, tone, and the number of interruptions to match the social norms of the occasion. With the sheer number of alarms and alerts vying for people's attention each day, technology should aim to match social expectations in a similar way and be more polite and contextually aware when seeking attention.

RAISING THE SITUATIONAL AWARENESS OF MACHINES

FIGURE 7.3
Google's Project Soli miniaturized radar so that devices could be more aware of what was happening around them.

A radar-based technology that detected nearby movement and gestures was once embedded in Google's Pixel phone, giving it significantly better situational awareness. According to the Google AI Blog, this technology, called *Soli* (shown in Figure 7.3), created an invisible bubble of radar signals around the phone, and movement within that bubble triggered actions such as "priming the camera to provide a seamless face unlock experience, politely lowering the volume of a ringing alarm as you reach to dismiss it, or turning off the display to save power when you're no longer near the device." Additionally, it allowed users to control the phone through touchless gestures.

Soli was only included in the 2019 Pixel 4 phone, likely removed from subsequent versions to reduce hardware costs and extend battery life, but it reappeared integrated into the Nest Hub in 2021, packaged as a way to monitor your sleep quality at night by detecting your movement without having to have a camera-based device in your bedroom—something most consumers are understandably reluctant to allow. With the growing need for devices to be smarter about what's happening in the environment around them and the coronavirus pandemic highlighting the need for touchless, gesture-based interfaces in public spaces as well, it's highly likely that the technology powering Project Soli will continue to show up in a variety of Google products for years to come.

How to Grab Attention

Human factors researchers know quite a lot about how humans process alarms and other auditory signals and how those signals can work in conflict or harmony. Here are some practical considerations from Christopher Wickens, Sallie Gordon, and Yili Liu's 1997 classic *An Introduction to Human Factors Engineering*:

- "The alarm should be tailored to be at least 15db above the threshold of hearing above the [background] noise level, 30db difference will guarantee detection."

- "Low-pitch sounds mask high-pitch sounds more than the converse"; therefore, "it is also wise to include components of the alarm at several different frequencies well distributed across the spectrum." This practice reduces the risk that an alarm will get masked by another alarm or sound in the environment and, if used smartly, can allow certain alarms to be delivered at lower overall volume while still cutting through.

- "We are generally better at dividing our attention between one visual and one auditory input than between two visual or two auditory channels." And "even when the same resources are used for two tasks, the amount of interference between them will be increased or decreased by differences in the similarity of the information that is processed." Therefore, several beep-based alarms all going off at the same time will make it extremely difficult to distinguish between different alarms. And while a tone alarm plus a speech-based alarm may be easier to distinguish from each other, the speech-based alarm may interfere with speech communication between the humans who are attempting to deal with the emergency.

Since Wickens et al. laid out these rules in the 1990s, human capacity for taking in alarms hasn't changed a bit, but technology has come a long way. As a designer, you can now use the microphones, cameras, and other sensors built in to devices as a matter of course to create alarm strategies that are context aware. To start, instead of setting an alarm at a standard 30db higher than the average sound level of an average environment to "guarantee detection," any system with a microphone might instead listen to its environment and play an alert at only 15–20db louder than the detected surrounding sound level. Additionally, cameras, wearables, and motion sensors built in to devices and architecture allow systems to be more aware of where the human is in proximity to the speaker that is sounding the alarm.

Like the Canary Box mentioned previously, the volume level for nonessential sounds should make way for more critical auditory information. Some cars use this trick already. In some cars, when you place your vehicle in reverse to back into a garage, the volume of the radio might be halved to allow you to hear noises in your surroundings better, as well as the auditory signals from proximity warning alerts.

These sorts of cooperative, hierarchical trade-offs between different systems is undoubtedly easier within a single vehicle, but it's not as impossible to achieve in an environment with devices from multiple manufacturers as you might think. Edworthy offers this idea to coordinate competing alarms coming from multiple devices within a single operating room (which is a chronic problem). "You could manufacture a universal box where you collect pieces of equipment and you could impose an alarm system on any [machine]." So, instead of each piece of equipment blaring out its own alarm in competition for the user's attention, they could all plug into a single source that appropriately prioritizes and communicates the status of all systems in a balanced, orderly fashion.

Maximizing Meaning in Alarms

"The alarm should be informative, signaling to the listener the nature of the emergency and, ideally, some indication of the appropriate action to take," according to Wickens et al.

Prior to the electronic age, creators of alarm systems were limited by the available technology of their time. Banging bells was probably the earliest form of large-scale alarm design—bell use has been traced back to the third Millennia BC among the Yangshao culture of Neolithic China. In the 18th century, there was the rise of more air-based horns and sirens, which brought such iconic sounds as the "ah-ROOOO-gah" sound of the klaxon horn and the spooky rise and fall of the WWI hand-cranked air raid sirens. But do these alarm sounds carry intuitive meaning? Are they "informative"? In a way, yes.

Most people who are exposed to Western media have intuitive associations with a handful of alarm noises. The ones mentioned above, as well as the wail of an ambulance, the startling blasts and whirls of an American-style police siren (a standard for most countries outside the EU), and the slightly more pleasantly trumpeted, dual-toned, alternating European-style police sirens. All of these various warning sirens and horns tune to minor musical intervals to convey a tense tone and sense of urgency. If they were tuned to a major interval, such alarms might seem to signal the arrival of something good like a herald trumpeting the entrance of royalty—something to look forward to rather than something to look out for. In contrast to these alarm sounds, the tones that sound when you arrive at a new subway station vary greatly in cities around the world; however, they are nearly always set at major intervals to evoke a positive association and a feeling that, "You have arrived!"

When it comes to the ubiquitous electronic "beep" of modern device alarms, that sound is so widely used for so many purposes, it has far less intuitive associations, with one major exception: the heartbeat flatlining sound effect. This sound is notable because it just uses rhythm, not melody, to convey meaning, "beep...beep...beep... beeeeeeeeeeeeeeeeeeeep." However, its frequent use in media has ingrained its deadly meaning in our psyche.

But beyond that particular use, beep-based alarms have very little intuitive meaning. The electronic beep first entered popular media with the launch of Russia's Sputnik 1, the first artificial Earth satellite, in 1957. The wall-to-wall press coverage of this achievement was the first time many people around the world had ever heard an electronic beep, but it would soon pervade modern life.

Today, there are tiny, affordable speakers and microchips that provide an opportunity for designers to significantly expand their "vocabulary" for alarms and alert noises to nearly anything they can imagine. However, so many of them still fall back to beeps, either because that is what is familiar or because these noises are required through regulations. But with the introduction of hi-fidelity sound systems into digital devices, designers have the opportunity to pack so much more information into the sounds that emerge from the machines.

One area of opportunity, of course, is the introduction of speech, and in some circumstances, this may be the ideal solution. However, for many reasons highlighted in the previous section, there are issues of conflicting sounds that can often make speech a less-than-ideal solution. Additionally, speech-based signals tend to be slower to deliver and slower to process. And, of course, there is the issue of choosing a language—there is no one language that is universally understood.

Between the vagueness of beeps and specificity of speech lies the inherent, intuitive meaning of everyday sounds. People from all cultures can easily identify certain sounds like the dripping of water, a child's laughter, or the sound of wood breaking. Chapter 3, "Intuitive Assessment," discussed using visual metaphors to help someone intuitively understand how something works, such as putting the plug for the electric vehicle in the spot where someone would normally find the opening of the gas tank. There is a lot of interesting exploration happening right now in using a metaphor to deliver intuitive messages beyond visual designs.

Sound Icons

Some of the most interesting and transformative work that Dr. Judy Edworthy and her dream team are doing is developing a revolutionary set of sound icons (also called *auditory icons* or *earcons*) for use in clinical settings. These sound icons use metaphor and layered sound effects to pack a ton of information into an extremely quick signal.

They begin with sounds that have a strong metaphorical association with the warning that is being conveyed. For example, the sound of a pill bottle being shaken is used to convey an issue with medication, the sound of a tea kettle is used to signal that a patient's temperature has risen too high, the sound of a ventilator (imagine Darth Vader's breathing) signals issues with respiration, and the classic "thump thump, thump thump" of a heartbeat signals, you guessed it, an issue with the heart.

These sound icons are layered on top of a second sound—a more classic attention-grabbing beep sequence. There are two versions of these beeps: a more low-key, warning-level beep pattern and a more insistent danger-level beep pattern. The sound frequencies of these beeps were carefully chosen so that they could play simultaneously with the sound icons without masking or conflicting with the icon. Both the warning beep and sound icon can be heard perfectly, even though they occur at the same time.

In simulated clinical settings, these sound icons are 26 times more recognizable than the beep-based alarms dictated by current standards, and they bring clinician response time down from 15 seconds on average to 12 seconds, a 20% improvement. This study and others point to a huge untapped potential in the use of auditory icons within the medical device category. And it's these sound icons that Edworthy and others are championing to be added to the international standards that regulate medical devices.

There are plenty of examples of sound icons in the consumer space—the crumpling of paper sound when you send something to the trash on a Mac computer, the whoosh of a sent email in Outlook, and the bird-like Twitter whistle come to mind. But these are mostly brand specific and rarely convey much more than "confirmation," "error," or "new message." There may be significant remaining potential in the consumer space to make sounds, especially alerts, more inherently meaningful, especially if designers begin to share some common auditory metaphors across platforms. Just as shared metaphors in graphic design—buttons, sliders, even the controversial "hamburger menu"—grow in number and become more intuitive as more brands use them, so too could common sound icons begin to widen the palette of sound designers.

Intuitive Touch

Haptic signals that convey information through vibration, force, touch, and movement have been in use by designers for hundreds of years. One of the earliest patented designs that used haptic feedback was a system used on steamships in the mid-1800s, which gave pilots the ability to steer better through the "feel" of their controls. Similar uses of haptic feedback were seen through a lot of the early (and current) implementations of the technology—creating an artificial feedback loop that mimicked the natural feedback someone would receive from the physical world if they were doing the task without the intervention of the technology.

But haptic signals can do more than just replicate natural physical feedback lost through automation. They can also be used to alert a user when something goes wrong. The "shaker stick" in a plane that vibrates the main steering control stick when a plane is tipped too far back and is about to stall is a classic example of this function.

The military is also experimenting with haptic-based navigation signals that give soldiers instructions on where to go through a vest they wear. This system is extremely quiet, hands-free, and eyes-free. Instead of staring at screens, which would reduce situational awareness, soldiers follow vibrations in the vest—turn left when they feel a vibration on their left side, turn right when they feel it on their right. They can also learn additional signals through memorizing patterns. But, of course, relying on communication through patterned buzzing is about as intuitive as communication that relies on someone learning a pattern of beeps.

Haptic designer Keith Kirkland is working to develop more intuitive haptic signals. "We call it a haptic language, but in a way it's not," Keith said. It takes years for most people to learn any kind of complex language. "I could give you 'buzz-buzz-buzz, buzz-buzz,' and you would eventually learn that that particular vibration means stop."

But as discussed previously, humans max out at about five to eight signals they can distinguish without an in-depth period of study. So Keith and his team decided that "using language mechanics wasn't the best way to deliver information." Instead, they turned to a much richer pool: metaphor. "Most of us have the experience of being in a vehicle and having that vehicle stop suddenly, along with our bodies' reaction to it. So if I wanted you to stop, I could send you a vibration pattern, sure. Or I could design a haptic experience that just made stopping feel like it was the right thing to do."

Keith is a cofounder, along with Kevin Yoo, of a company called *WearWorks* that designed a wrist-worn haptic device called *Wayband* that helps people navigate cities without the need to stare at a phone screen. This is a handy device for any user, but is especially appealing to users who are blind or visually impaired and can even be used by people who are both blind and deaf with some training. An early prototype of the Wayband helped a runner who was blind complete the first 15 miles of the New York City marathon without sighted assistance. (A sudden hard rain unfortunately shorted out the prototype, which had yet to be waterproofed.)

The band works by creating what Keith describes as a "haptic corridor." The band knows where the user wants to head next, based on their programmed route. "Based off the angle you are facing, relative to that point, we've designed a haptic experience that is 360 degrees. If you're facing the correct direction, you feel nothing. If you turn 10–20 degrees, you start to feel a tiny buzz. You turn back to 160 degrees, and you feel a huge buzz. At 160–180 degrees, which is the wrong way, you feel the harshest vibration we can deliver to you," Keith said.

Thus, 85% of users who put the band on can figure out which is the correct direction to head with zero instruction. (Keith said the success rate is nearly 100% for test users who are blind.) For such a unique product, that is an incredibly high rate of intuitive adoption. The sensation that the vibration-free area creates has been described by some users as a "window" or "open gate," which they intuitively understand they can walk through.

The original design team, who were all sighted, at first used a "follow-the-light" type of metaphor inadvertently. "We originally designed it so the strong vibration was the right way to go, but from feedback, we decided to reverse it. Because for people who are blind, it was just way clearer that you would walk toward nothing. It felt freer not to have to focus constantly on the signal to remain on the path." Not only did this improve the intuitiveness and enjoyability of the experience, but it also had major battery benefits as well. Users were facing the correct direction the vast majority of the time while using the product, so only activating the haptic motor to "nudge" the user back on track had the added bonus of extending the battery life significantly.

These insights came after the team embraced the adage, "Don't design *for* people with disabilities, design *with* them." WearWorks now employs consultants with visual impairments, bringing them to the table from the very beginning of development for new features and designs.

Keith and the WearWorks team continue to push at what's possible with intuitive haptic signals. They view haptics as an underutilized communication channel with huge potential. "With this proliferation of screen-based devices, you have an overdependency on the visual channel as a means of communication. You get all of your information through your screen and secondarily it's all audio after visual. So for people who don't have access to sight or don't have access to hearing, the world of design has become skewed."

With the buzzing haptic controls natively available within all mobile devices, every app designer already has the ability to incorporate simple haptic signals into their designs. And while not all buzzing products can produce experiences as sophisticated as Wayband's "haptic corridor," as wearables become more common and haptic motors become more refined, this will widen the palette even further for designers. It will be exciting to see what patterns and metaphors emerge as the communities of haptic designers grow and play off each other.

Critical Information: Alarms and Alerts

An overabundance of alarms and alerts used in modern electronics is creating a soundscape of constant noise and interruptions. The constant din has been proven to elevate stress, damage hearing, and have measurable health impacts. The constant ringing of alarms has become especially egregious in hospitals, and alarm fatigue has been tied directly to hundreds of deaths. However, innovative design and research done in the last few years point to some promising techniques for better alarm strategies.

How can designers create systems that communicate status changes calmly yet clearly?

Instead of creating systems that constantly interrupt just to communicate that they are working or to convey insignificant changes, consider lowering the fidelity of the information to create less disruptive updates and more opportunities for creative visualization, such as music that uses major keys, minor keys, and dissonance to help reassure people when things are going right and draw increasing attention when things start to go wrong. Additionally, in environments where mode confusion is common among humans and automated systems, designers can create highly prominent visual indicators. When mode changes occur, they can try to augment visual indicators with auditory warnings, especially if the automated system is disengaging unexpectedly.

What techniques can designers employ to make alarms and alerts smarter and more context aware?

Designers can use sensors, like cameras, motion detectors, and microphones, to monitor the sound levels and other environmental factors that would affect someone's interaction with an alarm and adjust sound levels and timing to allow better prioritization of competing alerts. In an environment with multiple alarm systems, they can consider ways to merge the data streams and create a central prioritization system for a cohesive alarm strategy.

How can designers maximize the meaning intuitively conveyed through each alarm signal?

It's important for designers to utilize metaphors to strike a balance between the vagueness of beeping alarms and the specificity of speech-based alarms. Metaphor-based sound icons and haptic signals are learned more quickly and prompt faster reaction times in users.

Go to the Source

99% Invisible: "Sound and Health: Hospitals": Part of an excellent two-part podcast series hosted by Roman Mars, 2019.

"Alarm Fatigue: A Patient Safety Concern": A study detailing the extent of the alarm fatigue issue by Sue Sendelbach RN and Marjorie Funk RN, 2013.

"To Reduce Hospital Noise, Researchers Create Alarms That Whistle and Sing": An article by Emily S. Rueb in the *New York Times*, 2019.

"Wearable Tech That Helps You Navigate by Touch"—TED Talk: A talk about intuitive haptic signals by Keith Kirkland, 2018.

"How Does A Siren Work?" (Mr. Wizard)—YouTube: A quick explanation of how air-based sirens work from a classic kid's science show, 1984.

Calm Technology: A book about nonintrusive design by Amber Case, 2015.

"'Mode Confusion' Vexes Drivers, Carmakers": An article by Junko Yoshida from *Electronic Engineering Times* that contains the full interview with Missy Cummings, 2021.

Hero by Design

What does it mean to be "good in a crisis"? Typically, we say this about people when they keep a level head during a stressful event, neither succumbing to fight-or-flight-driven hysteria, nor freezing up in indecision. People earn this label by quickly identifying what needs to be done and executing those actions competently. If they successfully save the day, then we might even call them *heroes*.

Society used to view heroes as highly unique individuals with some internal quality that they were born with that others did not have. But modern-day psychologists have rejected this idea. Just as the capacity for evil acts lies within each of us, so does the capacity for heroic acts. Stanford psychology professor emeritus Philip Zimbardo has made a study of the "banality of heroism" and believes that heroism is a much more common trait than we've been led to believe. His study in this area has led him to believe that if people think of themselves as heroes in waiting, if they practice stepping up and doing the right thing in the small moments of everyday life, they will be primed to take action in the big moments of crises as well.

Throughout this book, we've discussed ways to help users survive and thrive in all types of high-stress situations. Now we're going to bring it all together and talk about designing for those most critical moments when it's *your* users and *your* designs that will make the difference between disaster and saving the day. We're going to look closely at the parts of heroism that can be influenced by design and explore these key questions:

- How can designers spur users to step up to take on difficult challenges and persevere in the face of overwhelming odds?

- How can designers best support users in their valiant efforts to save the day?

- What are the human factors that go into a heroic intervention and how can design unlock them?

Inspiring Perseverance

When the Perseverance rover was sent to Mars in July of 2020, it left behind its twin, OPTIMISM, a high-fidelity test unit (shown in Figure 8.1). Because it's impossible to predict every challenge that Perseverance would run into on the red planet before sending it there, NASA created OPTIMISM to allow their team to test and troubleshoot possible solutions to the problems Perseverance might run into, even when it was hundreds of millions of miles away.

FIGURE 8.1
OPTIMISM, a test unit replica of the Perseverance rover.

The names of these rovers are especially well paired, because in life, optimism is critical to keeping our perseverance running in the face of unexpected challenges. In his study of human performance in difficult situations, human factors expert James Reason (mentioned in Chapter 2, "The Startle Reflex") found this need for optimism to be especially important when stress levels rise. He shared a story of a doctor performing a complex heart surgery on a newborn. The child had to be placed on a heart and lung bypass machine six times during the operation due to continued instability. However, the surgeon remained confident that he would be able to solve the problem and kept coming up with new solutions and trying them until the surgery was successful. Reason credited the surgeon's calm optimism for allowing his mind to remain flexible, "At no time did he show any outward signs of stress, and he maintained a belief throughout that the problem was resolvable."

The most critical factor in someone's ability to step up and save the day is their *belief* that they can. Stanford-based psychologist Albert Bandura calls this *self-efficacy*, and it is often the single greatest predictor of success.

When someone has strong optimism and self-efficacy, they try more often and improve their chances at success through the sheer law of numbers. They are also more resilient in the face of failure because their high self-efficacy causes them to attribute failure to a lack of luck or hard work, rather than personal inadequacy. Those people with low self-efficacy tend to lose control of their stress response more quickly when they meet challenges, often developing tunnel vision and being unable to engage in creative problem solving. And those who completely lack a sense of self-efficacy fail automatically because they never even try to solve their problems. "Unless people believe they can produce desired effects by their actions, they have little incentive to act," Bandura explained.

NOTE SELF-EFFICACY VERSUS SELF-CONFIDENCE

While a healthy self-confidence can certainly increase someone's self-efficacy, there's more that goes into self-efficacy than just a judgment of one's own capabilities—for example, evaluating the surrounding circumstances is also a big factor. For instance, someone with high self-confidence but low self-efficacy might say, "I'm a truly amazing soccer player, but even I can't kick a goal against a 200-mile-an-hour wind."

Bandura found that someone's level of self-efficacy is primarily determined by their past experiences in similar situations, but it's also heavily influenced by how well or poorly they've seen other people perform the task, the encouragement (or discouragement) they've received from other people, and the "gut feelings" and other physiological signals their body sends them, such as butterflies in their stomach.

While interfaces can't do much about those tummy butterflies, smart designs can absolutely take advantage of the research of Bandura, Reason, and others to build self-efficacy in users through these methods:

- Giving encouragement
- Showing what good looks like
- Highlighting available assets

Giving Encouragement

Other parts of the book have explored the psychological benefits of celebrating progress and giving clear, fast feedback. But another important opportunity to encourage your users comes as they are about to begin a difficult task, helping them overcome initial uncertainty or hesitancy to act and priming them to give their best effort.

A HEALTHY OPTIMISM

George Bernard Shaw once wrote, "Reasonable people adapt themselves to the world. Unreasonable people attempt to adapt the world to themselves. All progress, therefore, depends on unreasonable people."

Until 20 or 30 years ago, psychologists believed that a healthy mind was one that had an accurate understanding of reality and operated free of self-delusion. But as writer and Hidden Brain podcast host Shankar Vedantam revealed in the episode "Useful Delusions," "People with depression and anxiety may actually be seeing the world more accurately than people who are considered 'mentally healthy.'" It's the harshly accurate view of the world, stripped of optimism, that leads to an inability to function in a typical way.

Being optimistic is, in fact, so prevalent in the general population, that despite the indisputable evidence that it chronically distorts perception, psychologists decided it could not be labeled as a disorder, for to do so would be to instantly call more than half the world's population mentally ill.

Zimbardo's research into enabling heroic action found significant evidence that simply telling someone they have the authority to step up in a moment of crisis will significantly increase the likelihood they will actually do so when the big moment comes. He tells this striking story:

> In 2008, there was a massive earthquake in China's Szechuan province. The ceiling fell down on a school, killing almost all the kids in it. This kid escaped, and as he was running away, he noticed two other kids struggling to get out. He ran back and saved them. He was later asked, "Why did you do that?" He replied, "I was the hall monitor! It was my duty, it was my job to look after my classmates!"

As you design systems, consider how you are positioning the role of your user. Are they told through labeling and layout that they are a passive user, or are they explicitly empowered to direct, control, and take command when needed?

There are, of course, more explicit ways that interfaces can prime a user for heroic action. There's something unique and powerful that happens in the human brain when someone you trust tells you, "I believe you can do this." Research by psychologist Matthew

Redmond suggests that, while encouragement from a manager or teacher nearly always has some positive effect on performance, the level of impact increases significantly if the person receiving the encouragement respects and trusts the person who is giving it. This dependency can limit how effective these sorts of messages can be coming from an interface instead of a real human. The message of "I believe in you" coming from a machine feels empty because we know machines don't hold beliefs.

For system-generated encouragement to be believable, it should be rooted in what people already trust machines to do well—tell facts. For example, a system might be able to remind someone that they've done well in similar situations in the past or how their past performance compares to others. Something like, "Your 82% success rate on tasks like this is well above average! Good luck on this next round, and keep up the great work!"

Video game designers have long experimented with ways to encourage players to keep going even when the going gets tough. One beautiful example of this comes from the 2018 game *Celeste* from game creator Maddy Thorson. As Anders Furze of *Daily Review* put it, "*Celeste* is inarguably the hardest game I've ever played, but it's also one of the few that I've determinedly finished." Thorson liberally encourages users from the first minute of play on with supportive messages like the one shown in Figure 8.2

FIGURE 8.2
Starting within the first minute of play, *Celeste* offers its players generous amounts of encouragement.

Additionally, *Celeste* proactively offers ways to modify the level of difficulty. "We understand that every player is different," a nonjudgmental message states at the beginning of the game. "If *Celeste* is inaccessible to you due to its difficulty, we hope that Assist Mode will allow you to still enjoy it." These controls enable players to adjust the game's speed, grant themselves invincibility, or skip chapters of the game altogether. Modifications like these ensure that the player feels confident in their chance for success.

In addition to being able to actively adjust difficulty levels, the game is exceptionally kind when it comes to in-game death. "Respawning is just as quick as dying, with barely enough time for a Mega Man-like sparkle to signify your death, and the checkpoints are smartly placed to be forgiving while still making you prove you can complete the challenge in front of you," said IGN reviewer Tom Marks.

The game tracks your death count as you play and prominently celebrates failure as an opportunity for learning, shown in Figure 8.3.

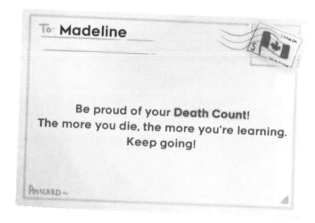

FIGURE 8.3
Embracing the principles of a growth mindset, *Celeste* actively celebrates how many times a player has died while trying to complete the game, pointing to it as proof of learning.

All of these resiliency focused interactions live in the story-driven themes of the game, such as the main character's struggle with depression and anxiety. "There aren't many games out there that feature two characters having a candid discussion about what depression feels like, or that depict a panic attack as tentacles attacking your character," Marks said.

Furze was equally taken with the interplay of story and game play, declaring *Celeste* the best video game of the year for the *Daily Review*.

"I finished this game a little wiser, a little more open, a little more confident in my own resilience."

Games certainly aren't the only interfaces where users experience challenges and need encouragement. Encouragement is essential for any type of training module or long, complicated flows, such as applications or tax filings. In fact *TurboTax* is a great model with encouraging language sprinkling liberally throughout their experience with headlines like "You're speeding toward the finish line!" Whatever type of process you are designing, if it represents a significant challenge for your users, don't skimp on the words of support. They might be just what your users need to meet the challenge.

Showing What Good Looks Like

YouTube's vast library of DIY videos is a great example of digital content that helps users build a sense of self-efficacy by allowing people to see another person model what success looks like. Video-based learning is so empowering, effective, and popular that a Pew Research survey found that 51% of YouTube's 800 million users use it as a go-to source to learn how to do something new. Additionally, extensive studies within the field of education have shown videos to be a highly effective learning vehicle, especially when videos are clear, short, engaging, and interactive.

When it comes to building confidence and a sense of self-efficacy, studies show that the human element in these how-to demonstrations is a critical component. It's not just the information in the video that builds someone's confidence, but it's actually seeing someone else do it successfully that really boosts your self-assurance. And, in fact, through a series of studies, University of Chicago researchers showed that the more often folks watch a DIY video, the higher their confidence grows. For instance, when asked to estimate their score in a dart game, participants who watched a how-to video about dart throwing 20 times picked an estimated score significantly higher than participants who only watched the video once.

However, it should be noted that watching the video 20 times did not actually improve their performance more than watching the video once. The same was found to be true for learning how to moonwalk, juggle bowling pins, or snatch a table cloth out from under a table setting of dishes. Watching someone else do these things over and over greatly improved confidence, but not actual performance.

Both Bandura's self-efficacy work and the Harvard studies speak to how strongly people's brains are wired to gain confidence through observation of other humans doing the thing they want to do. The Harvard studies might be seen as a cautionary tale about overconfidence, but overconfidence could also be said to be an evolutionary necessity. Failure is something that humans usually avoid at all costs—it's embarrassing, awkward, and sometimes painful. But trying and failing are a necessary part of achieving almost anything in life. Having an unrealistically high estimate of your chances of success is sometimes the only thing that actually gives you enough confidence to overcome your initial reticence. There are also many situations, especially those that require "heroic" intervention, where throwing a "Hail Mary" pass that has only a small chance of getting to the goal is a better option than trying nothing at all and having failure be assured. In these circumstances, having an accurate gauge of one's ability is less important than being willing to try.

If you design a system where it's in your users' best interests to try even if they might fail (and that's usually the case), but those same users are hesitant to take a reasonable risk, consider ways to integrate videos and images modeling what "good" looks like. For instance, say that you had designed a self-check-in kiosk at a hotel or airport lobby that no one seemed to be willing to use, even when lines were long. Placing a looping video on the screen of that kiosk showing someone using it to easily complete their check-in might be just what your user needs to step up and try the process themselves.

Highlighting Available Assets

Imagine that you were stuck at the bottom of a deep stone well. Unless you are some kind of champion rock climber, your belief that you could climb up the wall to freedom would likely be pretty low—you probably wouldn't even bother trying. Now imagine that you spot a rope ladder hanging down the side. Suddenly, your belief in your chances of survival spikes, and you're inspired to make an effort to get out of the well.

Context and access to resources matter when it comes to self-efficacy. Your belief in your own success is not static; it changes fluidly as you learn more about a situation and as you discover new assets and resources or blockers. It's why one of the first things that nature survivalists will tell you to do if you find yourself stranded in the wilderness is to take a thorough stock of all your resources, both on your person and in your immediate surroundings.

Similarly, when it comes to creating an interface that boosts someone's sense of optimism, it is important that the access to resources isn't hidden away. It could mean prominently displaying available resources, or it might involve creating a strong information scent toward them. Self-efficacy falls dramatically when there isn't a strong information scent available for the assets your user needs. For example, I drove around a car that had the time incorrect for six months out of every year because I had no idea how to change the time for daylight savings, and there just wasn't a strong enough information scent within the car menu for me to ever even try to track it down. I didn't know where to start so I just... didn't.

One of the most valuable assets that your system can provide for a user is access to help. Just letting them know help is available can give a huge boost to optimism and self-efficacy. *TurboTax* does this exceptionally well, reassuring users they will be supported all along the way of the tax-filing process in a very human tone, like the screen shown in Figure 8.4.

FIGURE 8.4

This question at the beginning of the *TurboTax* experience establishes early the tone of empathy that continues throughout the process.

In Figure 8.5, you'll see clear paths in the upper right-hand corner of the screen for users to access help via search, chat, or face-to-face conversation with an expert (requiring a paid upgrade) from anywhere within the flow.

FIGURE 8.5
Accessing additional help within the *TurboTax* experience is intuitive and omnipresent.

When filing 2020 taxes, many people filed unemployment earnings from pandemic-related layoffs. *TurboTax* included special videos and screens for the 2020 tax year, reassuring users that they were not alone and taking extra time to explain how coronavirus-related changes, like working from home, would affect taxes. And they told users this help was coming right at the beginning of the experience, as shown in Figure 8.6.

FIGURE 8.6
The 2020 tax year involved unusual circumstances for many, so *TurboTax* added additional touchpoints to reassure users along the way.

One final consideration when highlighting available assets is how you are framing facts and figures. Compare these two statements:

"The surgery has a 90% survival rate."

"The patient has a 10% chance of dying during surgery."

Most people have a predominately positive, hopeful emotional response to the first statement and a predominately pessimistic, fearful response to the second, even though, statistically speaking, both statements are conveying the exact same fact. As discussed in Chapter 5, "Reasoned Reaction," your mind gives more weight to information in front of you and discounts the relevance of information you can't see. So if the survival number is presented without the negative number, the mind will place the emphasis on the hope and not the fear.

How might this mental trick be used in your own work to help your users remain optimistic? Consider ways to put the visual focus on the information that gives hope rather than the one that inspires fear.

It's not always an easy task because the mind does not always weigh each point of a gain or loss equally. If you create an interface that emphasizes how much of a resource is left, you may believe you are helping your user focus on the positive. Let's say your user has an hour to complete a task. You put up a clock showing the hour and begin a countdown. As the first three seconds tick off, the clock still shows 59 minutes and 57 seconds. Plenty of time! But imagine the emotional state of the user at the end of the hour, how panic-inducing it is to watch the last 3 seconds run out on a clock. The loss of the first 3 seconds is weighted completely differently in the user's mind than the loss of the last 3 seconds.

Each information display needs to be considered carefully within the context of the specific design you are creating and the risks involved for your user. It's up to you as the designer to determine which end of the scale will elicit the more visceral reaction in the context of your product and organize your displays to minimize the psychological stress and keep hope alive for your users.

Supporting Users in the Moment of Truth

Just picture it: the hero slides across the hood of her super cool, high-tech vehicle, ducks into the driver's seat, buckles in smoothly, and puts her hands on the wheel. The music reaches a crescendo as the camera closes in on her face. She pulls her sunglasses down and says in a dramatic voice, "Let's hit it."

But instead of zooming off to save the day, there's a mechanical cough, a sputter, and then the engine powers down with a mournful whine. The sci-fi dashboard flickers off. From the dark, she says, "Well, crap."

When you create designs used by people who are trying to take heroic action, it's not enough to inspire them to greatness—your design has to help them actually save the day. This whole book has been about helping your users do better under stress, but let's take a look at the tricks and techniques gathered from the previous chapters that are most important to keep in mind when designing for that moment when your user steps up to face a crisis head on:

- Harness the benefits of the stress response.
- Take advantage of intuition.
- Give clear direction.
- Clear away distractions and focus on what's important.
- Connect them to others.
- Help them practice a heroic mindset every day.

Harness the Benefits of the Stress Response

As discussed in Chapter 2, an acute stress response gives users faster reflexes and super-powered gross motor skills. To take full advantage of the physical benefits that come with an adrenaline rush, use physical buttons, knobs, and other analog controls whenever possible for tasks that are critical for avoiding a crisis. If physical buttons aren't an option, then enlarge digital buttons and tap targets to accommodate shaking fingers and impulse variability. And whether the controls are physical or digital, ensure that they are within easy reach at all times throughout the experience, especially in moments with a high likelihood for trouble.

Take Advantage of Intuition

One of the biggest benefits of intuition in a crisis situation is the way that it allows people to automate common tasks and decisions so they can focus their conscious efforts on tasks that require critical thinking. In an emergency, your user will need all their wits about them to focus on the crisis at hand. In Chapter 3, "Intuitive Assessment," you learned that interfaces can tap into your user's intuition by following well-established design patterns and not pulling focus or interrupting the user unless absolutely necessary.

Give Clear Direction

One of the things that most often prevents people from acting in a crisis is they aren't sure what to do. They worry they will make a bad situation worse, and they freeze up, as discussed in Chapter 4, "Fight, Flight, or Freeze." Interfaces can help users overcome the freeze response by giving clear direction and a singular, step-by-step course of action.

Clear Away Distractions and Focus on What's Important

In a moment of crisis, focus is critical. Just as discussed in Chapter 5, you should ensure that your interfaces are supporting users by giving them all the information they need and none that they don't.

Connect Them to Others

Don't underestimate the power of human connection during crisis. You learned in Chapter 6, "Recovery," that people are braver, more rational, and more resilient in the face of trauma when they know they are not alone. Sometimes, enabling a connection to another human being is one of the best things that your UI can do when the going gets tough.

Help Them Practice a Heroic Mindset Every Day

Throughout this book, you've read that in a moment of high stress, users fall back on their instincts and intuition. But it's important to recognize that the things that form intuition are exposure, repetition, and habit. So the most effective way you can help someone be a hero in a moment of crisis is to help them practice being awesome every single day.

I gave this book the subtitle "What Life-Saving Technology Can Teach Everyday UX Designers," but what I learned while writing it is that it's the everyday designs that influence how users ultimately behave when lives are on the line. No matter what you design, strive to create systems that help your users form good habits about facing the fear of failure, helping one another, and making good choices.

It's practicing these heroic behaviors every single day that truly prepares them to come through for themselves and others in a life-and-death situation.

The Human Factors of Heroics

In his book *The Human Contribution*, James Reason explores all the ways that humans contribute to causing catastrophes, but he also acknowledges that there are many instances where exceptional human actions stop a disaster from occurring.

He has identified four make-or-break factors of human heroics:

- "Sheer unadulterated professionalism"
- "Training, discipline, and leadership"
- "Inspired improvisation"
- "Luck and skill"

Each of these characteristics can be seen at play within the dramatic tale of NASA's Apollo 13 mission, long used as a juicy case study for human factors research. On April 11, 1970, NASA launched what was meant to be its third voyage to the moon. "Houston, we've got a problem" were the famous words astronaut Jim Lovell said to Mission Control after a liquid oxygen tank exploded during the crew's third day in space. What followed was the stuff of summer blockbuster movies.

Professionalism/Discipline

In the moments following the explosion, the first priority was to secure the immediate safety of the astronauts. With major damage to their oxygen supply, there was no time to waste. They needed to take immediate action, and they had to make exactly the right moves in order for the astronauts to survive. NASA had worked for years to ensure that they had a documented plan for any eventuality, and this would be a crucial test of their professionalism and preparation.

The Apollo training team had indeed run a simulation with a similar system failure a year prior. "The simulation finished with a dead crew—a virtual crew, but dead nonetheless," Reason wrote. For months afterward, the lunar module controllers ran test after test until they had found solutions for every possible variation of that failure scenario that they could imagine. They had then carefully logged and recorded each problem and related solution. Due to this professional, disciplined approach, when the tank exploded during the real mission, the engineers knew just what to do—move the crew into the lunar module and use it as a life raft, allowing the crew to reserve the few remaining oxygen and power cells in the control module for final descent back to Earth. And it was a good thing that the solution was on hand and the plan was able to be executed so efficiently.

"When the lunar module was powered up sufficiently to sustain life, there were just 15 minutes of life left in the last fuel cell aboard the Odyssey," Reason said. The NASA engineers didn't just think through every scenario, they also painstakingly tested, documented, and filed the solutions in ways that were easy to retrieve in the moment of need.

Obviously, the heroic human factors of discipline and professionalism were critical in this particular moment of the Apollo 13 mission, but were there *designs* that helped to bring these qualities forward?

If you look carefully, in the background of photos of NASA's Mission Control from this era, like the one shown in Figure 8.7, you can often catch a glimpse of shelves full of binders. It was these binders that contained the detailed, meticulously indexed, and cross-referenced contingency plans. When that tank exploded, the engineers were able to literally pull the solution off the shelf. It was the layout and design of the pages within those binders, the information architecture of the index, even the placement of the shelves that held the binders, that allowed the engineers to retrieve the information efficiently in a situation where every second counted. All of those design choices contributed to the success of this mission.

In the digital age, the equivalent structures to these binders would be content management systems and search engines. Well-designed data entry tools help to standardize data and clean it up as it gets entered. These sorts of tools enable professional practices as they reinforce processes and a disciplined approach. Additionally, search tools that help users put their hands on the information they need in the moment they need it can be critical factors in a crisis, especially if they gracefully accommodate misspellings and typos—the sort of clumsy slips that often occur more frequently as tension rises. These kinds of systems may seem mundane, even dull, but in a crisis, they are critical enablers, turning years of hard work and preparation into a single shining moment where the user has exactly what they need in the exact moment they need it. How can you design your product in such a way to ensure that the hard work of your user, in the past, pays off in a critical moment in the present?

COURTESY OF NASA

FIGURE 8.7
NASA's Mission Control during Apollo 13 Mission with binders visible in the back.

Improvisation

The next problem to solve in the Apollo 13 mission required some real ingenuity. The lunar module had been designed to accommodate two men for a day and a half, not three men for three days. The engineering team realized almost immediately that the carbon dioxide filters in the lunar module were not up to the task. The carbon dioxide levels would reach deadly levels long before the crew made it back to Earth. They had extra filters from the command module, but the lunar module filters were square shaped and the command module filters were round. In my favorite scene from the 1995 *Apollo 13* movie, the head engineer dramatically dumped a box of spaceship supplies on a conference table and instructed his team to figure out how to "fit a square peg in a round hole" using nothing but the supplies on the table (see Figure 8.8).

FIGURE 8.8
In this iconic scene from the 1995 film *Apollo 13*, all the supplies available in the ship are dumped on a table in Mission Control.

The surrounding group of engineers began grabbing at supplies like they were contestants on some kind of *America's Next Top Engineer* reality show starting an epic hackathon.

Ed Smylie, the actual life-support system team leader, said in later interviews that there was no dramatic supply box dump, but his team did work for two days straight to hack together a filter converter from available materials. The final version, as described by Reason, was made of "a sock, a plastic bag, the cover of a flight manual, and lots of duct tape." The prototype version built by Mission Control can be seen in Figure 8.9.

FIGURE 8.9
Deke Slayton (check jacket), Director of Flight Crew Operations, reviews the proposed filter adapter design with a group of other NASA directors in Mission Control during the Apollo 13 mission.

After Mission Control walked the astronauts through building and installing a version of the converter, carbon dioxide levels quickly returned to a safe level. The mechanism built by the crew, shown in Figure 8.10, worked flawlessly for the rest of the trip—a perfect example of "inspired improvisation." But what enabled this type of improvisation and how can designers facilitate similar behavior for users of their own systems?

FIGURE 8.10
A picture of the actual filter taken by the crew. It was destroyed along with the lunar module as part of the planned reentry sequence.

In Chapter 3, "Intuitive Assessment," we talked extensively about Gary Klein's research on expert intuition. Part of Klein's definition of a "true expert" is someone who can improvise. It's the ability to improvise, he said, that distinguishes experts from mere "pretenders." Those who are pretending "have mastered many procedures and tricks of the trade; their actions are smooth. They show many of the characteristics of expertise. However, if they are pushed outside the standard patterns, they cannot improvise." Obviously, Ed Smylie and his team of engineers had true expertise. They understood everything about how those supplies worked and were able to hack items that were never meant to work together into a passable solution.

To develop this sort of adaptability within your own users, you must ensure that they understand not just the "what" but the "how" and "why" of the ways things work. This can be challenging if you work on complex digital products. The quote from Arthur C. Clark comes to mind, "Any sufficiently advanced technology is indistinguishable from magic." One of the things people love about technology is when it just works. But that can leave your users helpless when it just doesn't.

Consider how much your system reveals about how it works. How much does it teach users about what's going on behind the screens? If your system goes down, do your users know enough about the service your system was providing to get the critical tasks done in an emergency situation? Will your users recognize when a situation is outside of the capabilities of your system to handle? Especially if your system uses algorithms to make decisions or suggestions, consider revealing, in plain language, what goes into those algorithms and how they are weighted so your users can decide when it's best to use their recommendations.

Often, as a designer, you want to simplify the complexities within a system, hide away how the sausage is being made, and that can be helpful to a point, but when things don't go as planned, you still want your users to have the expertise and power to take heroic action and intervene effectively.

> **NOTE** **FUN FACT**
>
> Ed Smylie is quoted on the Wikipedia Duct Tape Page:
>
> Ed Smylie, who designed the scrubber modification in just two days, said later that he knew the problem was solvable when it was confirmed that duct tape was on the spacecraft: "I felt like we were home free," he said in 2005. "One thing a Southern boy will never say is, 'I don't think duct tape will fix it.'"

Luck Plus Skill

The final challenge the Apollo crew faced, reentry, would take skill and more than a little bit of luck. After swinging around the moon, the crew executed a series of engine firings to align their craft properly for the trip home. But something was off. They would discover later that a burst water vent was leaking into space, pushing the craft off course. If the reentry angle was off by even a single degree, the ship would burn up on reentry into the atmosphere.

Unfortunately, the navigation system was significantly impaired. Gene Kranz, the Lead Flight Director, explained the issue in an oral history with NASA, "This explosion that occurred had set a cloud of debris around the spacecraft and frozen particles of oxygen. And we'd normally navigate with stars," but now the stars were obscured from light glinting off the debris. "All we could see was the Sun, the Earth, and the Moon." It was Chuck Deiterich, an experienced flight controller who suggested an older alignment technique used back during the early Mercury flights that used points on the Earth, such as the point where night turned to day, as navigation markers.

In the near-freezing lunar module, the crew was cold, miserable, and exhausted after many sleepless days. But these astronauts were all experienced test pilots—they had proven skills in staying calm and performing exceptionally, even in suboptimal environments. And after being patiently talked through the revised alignment procedure from the Mission Control crew, they were able to execute the necessary course correction flawlessly, even using a method they hadn't trained with.

On April 17, six days after launch, the Apollo 13 crew returned safely to Earth, surviving one of the most dramatic incidents in NASA's history.

Reason points to the crew's skill as the critical factor of success in this final leg of their journey, and undoubtedly that human factor was a key component. From a design point of view, the elements that allowed that skill to be put to use were the redundant backup controls built into the spacecraft. Even with the impaired navigation system, the ship was still flyable by someone with the right skills.

Today, redundancy continues to be a standard part of high-stakes engineering. For example, despite the fact that the SpaceX Crew Dragon is meant to function fully autonomously, it is still built with "two-fault tolerance," meaning that any two things can fail, including the autopilot, and the human crew can still get the ship home safely. But in modern consumer-facing applications, more often than not, the conventional wisdom says it is best to strip all redundancy out of systems in order to simplify and streamline them.

This idea, that in order to be seen as easy to use, a system must eliminate redundant controls, really picked up steam as the Mac operating systems began to diverge from the Windows systems in the 1990s. Windows had a clunkier UI for sure, but if things went wrong, you could get into the operating system settings, and, with some intermediate tech skills, you could often get things back on track. Apple computers went a different direction, creating a system that was simple, streamlined, and intuitively usable by a much wider range of users at the novice level, but the more technical tools were so difficult to use that it required expert intervention to correct nearly any issue. By removing these backup, redundant controls, it left users helpless when things went off track. As Apple's popularity rose, so did the design trend of simplifying interfaces and killing redundant controls in consumer applications. It's possible the trend has now swung too far in the wrong direction.

Perhaps designers should rethink the value of redundancy, especially if the systems they design enable actions that are mission critical to the user in that moment, like sending money to a loved one in a crisis or navigating to a hospital in a foreign city.

Giving users a single, streamlined path is great for the user when everything works as intended. But redundancy and more robust access are often the only things that allow a user to achieve success when something fails along the *happy path*, the process the designer intended the user to follow. What are the critical tasks that users turn to your product to complete? How many ways do users have to perform that task? To enable users to apply their skills in a critical moment, you must take steps to ensure that even if part of your system fails, your users still have alternative paths for putting that skill to good use.

NOTE TWO IS ONE

> In the U.S. military, they have a saying to describe certain essential objects: "Two is one, one is none." Troops are told they must carry at least two ways to make a fire, two weapons, etc. If there are life-and-death consequences for not having an item when you need it, then you'd better have brought a backup.

We all can't design for astronauts, but we all have the potential for designing for heroes. These human factors of heroism—professionalism, discipline, improvisation, luck, and skill—were enabled and enhanced by the systems, objects, and interfaces that surrounded the key players both on the Apollo 13 ship and in Mission Control. As a designer, never doubt that you have the power to unlock heroism in your users as well.

Critical Information: Hero by Design

According to Philip Zimbardo, the potential for heroic acts is not a special gift held by only a few, but something that lies within each of us. Under the right circumstances, anyone can be a hero, and good designs have the potential to help unlock those behaviors in your users.

How can designers empower users to step up to take on difficult challenges and persevere in the face of overwhelming odds?

Fostering a strong sense of self-efficacy (a belief in one's ability to be successful) is essential to inspiring someone to stand up and try to make a heroic intervention. And fostering a healthy optimism helps users keep trying even when the going gets tough. Designs can promote optimism and self-efficacy through these techniques:

- Give encouragement.
- Show what good looks like.
- Highlight available assets.

How can designers best support users in their valiant efforts to save the day?

All of the techniques for designing for stress discussed in this book can help users succeed in a critical moment, but these principles are especially important to keep in mind to enable heroic interventions:

- Harness the benefits of the stress response.
- Take advantage of intuition.
- Give clear direction.
- Clear away distractions and focus on what's important.
- Connect them to others.
- Help users practice doing the right thing every day.

What are the human factors that go into a heroic intervention?

According to James Reason, heroic saves often share these four elements:

- "Sheer unadulterated professionalism"
- "Training, discipline, and leadership"
- "Inspired improvisation"
- "Luck and skill"

Go to the Source

"The Psychology of Evil"—TED Talk: An inspiring talk on both good and evil by Philip Zimbardo.

The Human Contribution: Unsafe Acts, Accidents, and Heroic Recoveries: A book by James Reason, 2008.

Apollo 13: A Universal Pictures film directed by Ron Howard, 1995.

"Self-Efficacy: The Foundation of Agency": An essay by Albert Bandura found within the collection Control of Human Behavior, Mental Processes and Consciousness.

Useful Delusions: The Power & Paradox of the Self-Deceiving Brain: A book that explores the idea of optimism and other helpful biases by Shankar Vedantam, 2021.

"Easier Seen Than Done: Merely Watching Others Perform Can Foster an Illusion of Skill Acquisition": A study by University of Chicago researchers Michael Kardas and Ed O'Brien, 2018.

CODA

I'm backstage at a design conference trading tips for dealing with nerves with the other speakers. "I've got one I learned at THAT Conference," says Lauren Liss, referring to a family-friendly tech conference held at a giant indoor water park resort in Wisconsin. "About two hours before I have to go on stage, I go to the scariest water slide in the park where you get inside a tube and they drop you for a 50-foot free fall. It's completely terrifying. It dumps all of my adrenaline into my bloodstream at once. Then I get out, I go back to my room, take a shower, get gussied up, and by the time I walk out on stage I'm cool as a cucumber."

Now, that's a woman who knows how to hack a stress response, I think to myself, taking careful mental notes. There's something so empowering about the idea of taking control of your own fear that way—making it work for you.

So much of the human stress response is outside of the conscious control of the human who is experiencing it. But when it's well understood, and good design effectively applied, the response can be managed, even harnessed, to achieve the goals of the person who's experiencing it. It's been my hope in writing this book that more designers would learn to do just that.

One recurring theme that I hope you'll leverage to that end is the idea that intuition isn't mystical, it is simply pattern recognition. Intuition is a central driver of human behavior during a stress response. But, as a designer, you should never forget that you have the power to influence intuition. You can develop helpful intuition within the minds of your users through the introduction of new patterns, cementing them through repetition and training. You can also create systems that interrupt those times when an overreliance on intuition leads users astray, causing harmful bias.

Another theme that may help you anticipate stress-driven human behavior at the macro level is the idea that stress moves human abilities along a sliding scale with "primal" physical behavior on one side and "civilized" rational behavior on the other. Language, creativity, analysis, math, and logic, all of these more refined activities; they degrade quickly as stress levels rise—as do any activities involving fine-motor skills or detailed work. But the loss of those skills comes with a gain in raw physical power, an increase in strength, speed, focus, and awareness. Stress grounds people in the present moment, sometimes to the point of tunnel vision, and amplifies their "gut" feelings. In fact, intuition is so closely linked to people's physicality that intuitive thoughts often manifest themselves as physiological reactions, such as a sinking feeling in the stomach, raised goosebumps on arms, or a shiver down the spine. Identifying these overarching themes of moving from complex to simple, conceptual to literal, detailed to broad, cerebral to physical, etc., can help you make better design decisions when designing for users under stress.

And the final theme I hope you carry forward is the power of human connection in times of crisis. Getting to others, either to receive help or to provide help, is often a central driver of human behavior in times of crisis. Additionally, being with attachment figures has a protective effect during traumatic experiences, and connecting with a trusted person after a crisis is often one of the fastest ways to inspire a sense of calm. This need for human connection can be a difficult reality for digital designers who more often than not are designing experiences that are replacing human-to-human interactions with human-to-computer interfaces. But, the fact is, this is a true and important human need when something goes wrong in high-stakes situations and should be accommodated, especially in moments of anticipated peak stress.

No matter what you design, I hope you found something pertinent in this book and find new ways to look at old problems. I believe we, as designers, have a responsibility to create products that protect and empower the people who use them. However, it takes more than good intentions to achieve goals that lofty. Like any heroic intervention, it takes professionalism, discipline, improvisation, luck, and skill. It also takes humility, research, and the inclusion of many voices and diverse perspectives. I hope this book enables you to create designs that come through for people in the moments that matter most.

buttons and dials

    analog, in startle reflex, 31–35, 195

    design techniques for fight-or-flight, 99–100

bystander effect, 93

C

Cadillac's Super Cruise, 170–171

Calm Technology (Case), 165

calming aesthetics, in recovery period, 143–149

    curved vs. angular, 143–144

    nature and biophilic design, 146–148

    order and clarity, 144–146

    quick reference for design, 148–149

Canary Box, 172, 175

CareTunes, 166–168

cars. *See* vehicles

Case, Amber, 165

Cave, George, 66

Celeste video game, 188–190

chat boxes, and distractions, 89

chat icons, as escape route, 32

Chinese culture, compared with U.S. culture on values of people, 132

chip-reader payment machines, 74–75

choice paralysis (decision inertia), 93–95

chronic anxiety disorders, 25

chronic stress, 2

clarity, humans' preference for, 144–146

Clarke, Arthur C., 202

clinical trials, design for zero familiarity, 110–118

clutter, hiding, 122

cognitive bias, 52

Collinson, Sarah, 71–73

color contrast, 100

comic-style content, 113–118

comparative decision-making, 48

"Conditions for Intuitive Expertise: A Failure to Disagree" (Kahneman and Klein), 52

confidence, 190–191

confirmation bias, 52

Conflict (Shortland, Alison, and Moran), 94

conscious to unconscious control, 62, 137

consent disclosure forms, 111–113

content, design principles for zero familiarity, 113–118

content management systems, for professional practice, 198

contextual inquiry, 13, 124

contingency design, 42

contingency plans at NASA, in binders, 198–199

"contrast equals drama," 24

conversation and interpersonal skills training, 57–58

Cook County Hospital, 125–127

cooling down period, as de-escalation technique for UIs, 90–92

Coral Project, 90–91

core values, and least-worst decision-making, 94–95

coronavirus pandemic

    author's story of, xv–xvi

    and guidelines for rationing care, 128

    help in filing taxes, 193

    stress-driven activity, 151

cortisol, 5, 11, 12, 15, 140

CPR Womanikins, 72–74

Crandall, Beth, 60

crisis controls

    through voice control, 102–103

    for visual interfaces, 99–102

Critical Alarms Lab, 164, 166, 167–168

Critical Information. *See* summaries of critical information culture

    designing for morals and values, 127–132

    and mental pattern libraries, 50–52

Cummings, Missy, 169–170

Cunningham family (Landon, Shaun, and Ashley), 18–19

Current, Hunter, 68–69

curved objects, 143–144

customer journey map, 13

customer service, human support for users fleeing toward help, 86–87

D

data entry tools, for professional practice, 198

Daugherty Biddison, Lee, 129

Davis, Michael, 27–28

decision inertia (choice paralysis), 93–95

decision-making

    comparative, 48

    least-worst, 94–95

    naturalistic, 47–48

de-escalation techniques for UIs, 87–92

    acknowledgment of perceived injustice, 88–89

    cooling down period, 90–92

    details and follow-up questions, 89

    distractions minimized, 89

    plain language, 89

    tone, warm and respectful, 88

    we as a team, 90

design, purpose of this book, xvi–xvii

devil's chord, 167

Diabetes by Design posters, 155–156

Dietrich, Chuck, 203

digital buttons, lessons from analog, 31–35

digital interface design, and expert intuition, 53–54

directives, clear and specific, 93, 196

disassociation, 142

discipline, as factor in heroics, 197–198

dissonance in music, 167–168

distractions, 89, 196

distress, 2

diversity. *See also* biases

    importance of, xviii–xix

DIY videos, 190

drafts folders (digital feature), 42

duct tape, 200, 202

Duolingo, language learning, 58

Durussel-Baker, Alex, 154–157

E

earcons, 177

Edworthy, Judy, 163, 164, 175, 177

effective action, 149, 151, 152–153

ejection seat instructions, 108–109

"Emergency SOS" on iPhones, 97–99

emergency stop buttons, 31–34

    red color of, 32

encouragement, to inspire perseverance, 186–190

end of speech detection, 103

error. *See* human error

ethics in research, 13

eustress, 2, 25

execution failures, 40

The Exorcist (film), 26

experts, and intuitive assessment, 46–48, 52–54, 201

experts, design for reasoned reaction, 118–124

    appropriate task assignment, 119–120

    bias exposure, 120–121

    prioritize and organize, 121–122

    support the decision, 122–124

extrasensory perception (ESP), 46

impulse variability

    enlarged controls for visual interfaces, 99

    in startle reflex, 36–39

Indonesian design team, 50–51

injustice, perceived, 88–89

Instagram, Diabetes by Design posters, 155–156

insurance coverage story, 60–61

intelligence analysts

    expert intuition, 53–54

    experts in the unpredictable, 118–124

intensity, as factor in startle reflex, 20, 22

interactive learning, for intuition development, 61–62

internet use by world population, 53

An Introduction to Human Factors Engineering (Wickens, Gordon, and Liu), 174, 175

intuition development in users, 54–62

    feedback, clear and immediate, 58

    interactive learning, 61–62

    repetition plus variety, 56–58

    soldiers and training, 55–57

    stories and improvisation, 59–61

intuitive assessment, 45–77

    in Amy's accident story, 5

    bias and intuition, 70–75

    considerations, 7–8

    development of intuition in users, 54–62

    expert intuition, 52–54

    harnessing intuition, 62–67, 195–196

    length of, and brain region engaged, 14

    pattern's role in formation of intuition, 49–52, 207

    vs. reasoned reaction, 110, 119

    science of intuition, 46–48

    summary of critical information, 75–76

    superstition and intuition, 67–69

intuitive design, 49, 65–66

intuitive reading, 100–101

intuitive touch, 178–180

iPhone, Emergency SOS, 97–99

J

Jaws (film), 24

JOAN agency, 71–73

Johns Hopkins

    alarms in hospitals, 163

    guidelines for rationing care, 128–129

Johnson & Johnson Medical Devices, 64–65

Joint Commission, and alarm fatigue, 164

jump scares, as startle reflex, 19–20

K

Kahneman, Daniel

    on clarity and trustworthiness, 146

    intuitive assessment, 52–54, 71

    reasoned reactions, 119, 126

Kearns, Alexander, 80–81, 83–84

kerning, 101

Kirkland, Keith, 179–180

KISKA design firm, 66–67

Klein, Gary

    intuitive assessment, 46–47, 49, 52, 54, 60, 71, 201

    on least-worst decision-making, 94–95

    reasoned reactions, 119

knowledge-based mistakes, 40–41

Kranz, Gene, 203

Kubrick, Stanley, 23

L

language, designing for zero familiarity, 114–116

lapses of memory, 40–41

Nightingale, Florence, 147

Nilles, Nancy, 114–115

non sequitur, 20

novelty, as factor in startle reflex, 20

O

Olympics, starter pistol and startle reflex, 26–27

OmniAssist, 141

OnStar Emergency, 6

optimism, 185, 187, 194

Optimism rover test unit, 184–185

order, humans preference for, 144–145

Ozcan, Elif, 164, 166, 167

P

panic attack, in video game, 189

panic clicking, 33, 98

panic leading to suicide, 80–85

parasympathetic system, 11, 15, 136–138, 151

pattern recognition and categorization, and bias, 70–71

patterns, role in formation of intuition, 49–52

peak stress points, 12–13

Pedersen, Erik, 63

PELOD-2 (PEdiatric Logistic Organ Dysfunction) score, 130

perfection photos, 144–145

Performance Recovery Following Startle (study), 35–36

peripheral vision vs. central vision, and startle reflex, 21

perseverance, inspiring in users, 184–194

giving encouragement, 186–190

and optimism, 184–185, 187

and self-efficacy, 185–186

showing assets and resources, 191–194

showing what good looks like, 190–191

Perseverance rover, 184–185

phones, 911 emergency dialing, 96–99

pigeon experiment, 67–68

planning failures, 40

police

crisis negotiations, 88–89

and memory delay in trauma recovery, 142

sirens, 175

pop-ups, confirmation dialogs, 42

post-traumatic stress, 25

precision farming software, 68–69

predictable movements, in startle reflex, 27–29

prediction of human behavior, 54

prefrontal cortex, 2, 11, 15, 61

pretenders, 201

"primal" attributes, 3

priming, as factor in startle reflex, 24–26

professionalism, as factor in heroics, 197–198

professions, and expert intuition, 52–54

protection, in startle reflex, 28

Psycho (film), dissonant chords, 167

pull-over button, in taxi, 34–35

R

radar-based technology for situational awareness, 173

rage, 89

Rapier, Rhett, 64–65

ratings, in marketplace, 121, 123

Readability Group, 100

Reason, James, 40, 185, 197, 203, 205

reasoned reaction, 107–134

algorithms in medical decision-making, as protection against bias, 124–127

Virttex testing dome, 38–39

virtual reality training simulators, 56–58

vision, peripheral vision vs. central vision, 21

visual design for calm aesthetics, 148

visual interfaces, crisis controls for, 99–102

voice controls, 102–103. *See also* speech-based alarms

volume control
auditory triggers in video games, 22
physical dials in cars, 30

Vox Media, Coral Project, 90–91

W

Wayband, haptic device, 179–180

Waymo taxi, and pull-over button, 34–35

WearWorks, 179–180

Web Content Accessibility Guidelines, 100

Wickens, Christopher, 174, 175

Windows operating systems, and redundancy, 204

Winterrowd, Jeremy, 25

Wolf, Jackie, 111–113, 116–117

The Woman in Black (film), 20

Womanikins for CPR, 72–74

Woodhead, Muriel, 139

working memory, 63

World Trade Center, Ground Zero, 85–86, 150

World War II, separated children, 152

X

"X-ing out" of a website, 85

Z

zero familiarity, design principles for, 110–118
behavior modeling, 116–117
chunking information, 114
content, written and visual combined, 113
interactive information, 114
language, 114–116
relatable information, 117–118

Zimbardo, Philip, 184, 187, 205

Zuniga, Jorge, 124

ACKNOWLEDGMENTS

J ust as raising a child takes a village, so too does creating a book. I was honored, humbled, and delighted to discover just how talented and generous my own village is through the process of researching, writing, editing, and illustrating this book. There are so many people to whom I owe a debt of gratitude.

First and foremost, I must thank my family. My incredible husband, Jacob, took on so much to allow me the time to write. In addition to keeping our household running while I whiled away the weekends and evenings in front of my laptop, he provided unfaltering love, support, and insightful feedback all along the way, from my original proposal to my final illustrations. Without him, this book never would have happened. Also, I am exceedingly grateful to my amazing daughter, Olive, who was so supportive and understanding of her busy mommy. (Love you kiddo!) And I am also blessed by many other family members who encouraged me every step of the way, especially my parents, Jennie and Dave Hawkey, who were amazing cheerleaders, helpers, and shoulders to cry on, and my incredible in-laws, Laurie, Jim, and Andy Swindler!

I also was "in good hands" with my amazing Allstate UX coworkers who inspired, supported, and encouraged me throughout this project, providing valuable feedback, forwarding ideas and articles, and connecting me so generously to their networks. Special thanks are due to the following friends, colleagues, and leaders who enthusiastically supported my speaking and writing on the topic of life-and-death design: Lynn Schmitt, Chris Schoen, Laura Oxenfeld, Susan Thome, Tom Newsom, Daniel Nehring, Tom Dell'Aringa, Matt Hay, Sarah Bedoe, Brian Crowley, Daniela Rebic, Clif Simmons, Alicia Warren, Jamie Hsu, Jorge Zuniga, Erik Pedersen, Aisha Ghori Ozaki, Caitlin Everett, Gina Mete Onan, Nicholas Nottoli, Kristin Kowynia, Karlson Rapp, Maggie Hong, Leslie Chambers, Gregory Paczkowicz, Janice Nason, Roger Tye, Joanne Glennon, Trina Uzee, Naomi Minnich, and David Hernandez. I also want to send an extra special thank-you to fellow Rosenfeld Media author Michael Metts, for all of his support and coaching throughout the process and to Pradeep Nayar in whose memory this book is dedicated.

Throughout the creation of this book, I got to interview and connect with some truly inspiring designers and experts, many of whom you will meet in the pages of this book. These include Florian Gulden, Parrish Hanna, Josh Campbell, Jody Campbell, Alexandra Gaski, Keith Kirkland, Sarah Collinson, Jacob Metiva, Carrie Watcher, Jackie Wolf, Alexandra Gallant, Eva Penzeymoog, Melissa Hsu, George Cave, George Salazar, H Locke, Holly Schroeder, Alex Durussel-Baker, Andrew Losowsky, Becca Barish, Tamara Fox, Thai Dang, Brian Cugelman, Samantha Szymczak, Jennifer Strickland, Kelly Leonard, Eric Rodenbeck, Nicolette Hayes, Amira Hankin, Lisa Lipkin Balser, Nancy Nilles, Jenka Gurfinkel, Camille Gribbons, Rhett Rapier, Judy Edworthy, Joseph Schlesinger, Jon Bloom, Erin Null, Alan Hommerd-ing, Tim Grabacki, and Mark Pestana. The generous donation of time and expertise from each of these folks was critical to the success of this book. I also want to give a shout-out to the amazingly supportive folks in one of my favorite online communities ever, Space Hipsters, who helped me fact-check space stuff and dig up archival photos of every-thing from binder shelves to fighter plane manuals.

Another key group were my test readers and technical reviewers who helped me polish the book to a shine, including Lynn Schmitt, Melissa Smith, Erik Pedersen, Lauren Liss, Laura Oxenfeld, Jon Bloom, Sharon Ford, Carolyn Chandler, and Aisha Ghori Ozaki. Many of them also contributed kind words as testimonials to the cover of this book. A huge thank-you to Lisa Baskett as well, whom I admire so much. I was so thrilled and honored when she agreed to write the foreword.

And last, but certainly not least, there is the incredible team of con-summate professionals who made this book a reality. Marta Justak, my managing editor, has been an essential figure through this entire journey. I'm so glad she was working the Rosenfeld Media book mobile that fateful day we met, and I'm so glad I was able to work up the courage to walk over and say, "I've got this idea for a book…" What she told me that day was 100% correct, "Writing a book is hard," but with her support, guidance, and wisdom, I never doubted that it would get done and make it something we could all be proud of. Natalie Yee performed the inclusive language review for the book, and I can't thank her enough for her thoughtful comments and straightforward guidance. A big thank-you as well to Caitlin Everett and Shawn Morningstar, who both contributed illustrated figures to the book, to Bill Howard, Jesse Rieser, and Christopher Horner, who contributed photos, and to the dynamic duo, Brian Crowley and Wesley Wong, who created the incredible comic-book-style panels that bring so many of the stories in this book to life. Thank-you as well to Danielle Foster for her work on layout and to Adeline Crites-Moore, Karen Corbett, and the rest of the Rosenfeld Media team for your support throughout the project. And finally, a *huge* thank-you to Lou Rosenfeld for taking a chance on an unknown author and for all your guidance throughout the process.

Dear Reader,

Thanks very much for purchasing this book. There's a story behind it and every product we create at Rosenfeld Media.

Since the early 1990s, I've been a User Experience consultant, conference presenter, workshop instructor, and author. (I'm probably best-known for having cowritten *Information Architecture for the Web and Beyond*.) In each of these roles, I've been frustrated by the missed opportunities to apply UX principles and practices.

I started Rosenfeld Media in 2005 with the goal of publishing books whose design and development showed that a publisher could practice what it preached. Since then, we've expanded into producing industry-leading conferences and workshops. In all cases, UX has helped us create better, more successful products—just as you would expect. From employing user research to drive the design of our books and conference programs, to working closely with our conference speakers on their talks, to caring deeply about customer service, we practice what we preach every day.

Please visit **rosenfeldmedia.com** to learn more about our **conferences**, **workshops**, **free communities**, and **other great resources** that we've made for you. And send your ideas, suggestions, and concerns my way: louis@rosenfeldmedia.com

I'd love to hear from you, and I hope you enjoy the book!

Lou Rosenfeld,
Publisher

RECENT TITLES FROM ROSENFELD MEDIA

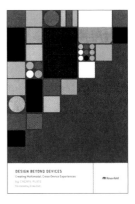

Get a great discount on a Rosenfeld Media book:
visit **rfld.me/deal** to learn more.

SELECTED TITLES FROM ROSENFELD MEDIA

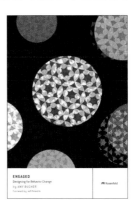

View our full catalog at **rosenfeldmedia.com/books**

ABOUT THE AUTHOR

Katie Swindler is a user experience strategist who writes and speaks on topics related to human-centered design. She was originally educated as a theater director, and because of that, she brings a unique perspective to digital work. She believes that if brands want to truly connect with consumers, they must combine emotion, utility, storytelling, and technology to fulfill real human needs.

Katie is currently the Design Strategist for Allstate Insurance's Innovation group, guiding the design of new products and services. Prior to joining Allstate, she was a UX Director at FCB Chicago, where she was the UX lead on the 2016 global redesign of JackDaniels.com, as well as the lead experience designer for many other clients such as Cox Communications, Meow Mix, and Toyota Financial Services. Katie is an experienced presenter who has spoken on UX topics internationally at design industry events such as SXSW Interactive and IxDA Interaction. She lives in Chicago with her husband, Jacob, daughter, Olive, and their snuggly pup, Rocket Rainbow.